Small, Short and Unsecured
Informal Rural Finance in India

Small, Short and Unsecured
Informal Rural Finance in India

F. J. A. BOUMAN

With the assistance of:

RENÉ BASTIAANSEN
HAN VAN DEN BOGAARD
HENNY GERNER
OTTO HOSPES
JOOST GROOT KORMELINK

DELHI
OXFORD UNIVERSITY PRESS
OXFORD NEW YORK
1989

Oxford University Press, Walton Street, Oxford OX2 6DP
New York Toronto
Delhi Bombay Calcutta Madras Karachi
Petaling Jaya Singapore Hong Kong Tokyo
Nairobi Dar es Salaam
Melbourne Auckland
and associates in
Berlin Ibadan

© Oxford University Press 1989

SBN 0 19 562454 8

Typeset at Taj Services Ltd., Noida
Printed at Rekha Printers Pvt. Ltd., New Delhi 110020
and published by S. K. Mookerjee, Oxford University Press
YMCA Library Building, Jai Singh Road, New Delhi 110001

Contents

Preface		ix
Introduction		1

1. **Formal and informal finance, a review of the debate** — 5

    1.1. Financial market composition — 5
    1.2. Formal bias — 6
    1.3. New interest in informal finance — 8
    1.4. Formal agendas for informal finance — 9

2. **Rural finance in India** — 11

    2.1. Main features of financial policy — 11
    2.2. Rural finance policy — 12
    2.3. Record of formal rural finance in brief — 12
    2.4. Nagging problems and weaknesses — 16
    2.5. Summary — 17

3. **The sugar boom in Maharashtra** — 20

    3.1. Cultivation of sugar-cane in Sangli District — 21
    3.2. Opportunity and response — 24
    3.3. Summary — 28

4. **The rural financial market in Sangli District outlined** — 30

5. **Co-operative Credit** — 34

    5.1. The Land Development Bank — 34
    5.2. District Central Co-operative Bank — 36
    5.3. Primary Agricultural Credit Societies — 39

CONTENTS

6. The Urban Credit Societies ... 44

7. The bishi, a self help organization of the informal finance sector ... 52

    7.1. Nature and origin of bishi: of ROSCA and RESCA ... 52
    7.2. Bishi emergence and rationale ... 56
    7.3. Comparing bishi and UCS ... 58
    7.4. Bishi operations ... 61
    7.5. Summary of bishi characteristics ... 65
    7.6. Recent developments in bishi culture ... 66

8. Of moneylenders and pawnbrokers ... 70

    8.1. Gold as everybody's piggybank ... 71
    8.2. Gold and pawnbroking in Asia ... 74
    8.3. The Indian experience ... 76
    8.4. Summary ... 80

9. Pawnbroking in Sangli District ... 81

    9.1. Pawnbroking by formal financial institutions ... 81
    9.2 Pawnbrokers in the informal market ... 86
        9.2.1. The licensed pawnbroker ... 87
        9.2.2. The unlicensed pawnbroker ... 92
    9.3. Pawnshop economy ... 97
        9.3.1. The licensed pawnbroker ... 97
        9.3.2. The unlicensed pawnbroker ... 98
        9.3.3. Institutional pawnbroking ... 98
    9.4. Summary ... 100

10. Dairy development and informal moneylending ... 103

    10.1. The private sector ... 103
    10.2. The State and the co-operative sector ... 105
    10.3. Milk collection and moneylending ... 108
    10.4 Summary ... 111

11. Evaluation and summary　113

　11.1. Finance and development in Sangli　113
　11.2. Opportunity and response of the financial sector:
　　　　planned growth versus organic growth　115
　11.3. Formal and informal finance: between myth and reality　117
　11.4. Formal and informal finance: substitution or
　　　　complementarity?　121
　11.5. Exploring new frontiers of financial technology　125

Notes　128
References　132
Appendix 1　137
Appendix 2　138
Index　141
Maps:　Map 1　Location of Sangli District in
　　　　　　　　Maharashtra, India　22
　　　　Map 2　Sangli District, subdistricts and
　　　　　　　　research villages　25
　　　　Map 3　Irrigation projects in the district　26

Preface

This book provides glimpses of the world of finance, both of the formal and informal kind, in the fast-developing rural district of Sangli in Maharashtra State, India. The emphasis is on informal financial intermediaries. Not all informal agents and institutions are discussed, for that would be nearly impossible, given the multitude and diversity of actors in the field.

Research for this study was carried out from 1984 through 1986. Local data collection was done by graduates of the Agricultural University of Wageningen, the Netherlands, in part-fulfillment of the requirements for a master's degree.

Attention focused first on *bishi*, an informal savings and loan society.* Then the focus shifted to the bishi's formal counterpart at village level, the Primary Agricultural Credit Society (PACS), that holds such a prominent place in India's rural finance policy. The Urban Credit Society (UCS) was also studied, as neither a completely formal nor completely informal financial institution. Contrary to the suggestion of its title, the UCS in Sangli is very much present in the villages in the district. Data on these three local institutions were collected by Henny Gerner, Joost Groot Kormelink and Otto Hospes, each staying in India for six months.

In 1986, the licensed and unlicensed moneylender and pawnbroker were included in the study, as well as the lending activities of the milk collector. Data on these were compiled by René Bastiaansen and Han van den Bogaard, who remained in the field for three months.

The study of the UCS is based on data from thirteen societies. Nine of these are located in six vilages in the (semi-) irrigated sugarbelt in Tasgoan and Walwa subdistricts and four in villages in dry Khanapur subdistrict. In these same communities, more than fifty bishis were in operation at the time of the study. Data were collected through interviews with members and office bearers. Monthly meetings of the eight bishis operating in

---

* *Bishi* is Marathi vernacular for 'money-matters', matters of a financial nature.

Vasagade in 1984 were attended and proved most informative. Discussions in government and university offices, in teahouses and farms alongside the road provided fresh insights. Further, twenty-two PACS were visited in these and other villages in Khanapur and the sugarbelt. This latter survey served the limited purpose of getting an impression of the way in which the lending business of an average PACS was conducted and of the economic viability of such a society.

To get a proper impression of the pawnbroking business, scores of banks, moneylenders and petty pawnbrokers were visited in villages and towns in both irrigated and dry areas of the district. The report on lending activities of the milk collectors took shape during a study of the development of the dairy industry in the district.

I supervised the research and visited the area myself twice in 1977 and 1985. I wish to thank the researchers for their contributions and enthusiasm; the data they collected and the impressions they gained are incorporated in this study. I, however, am solely responsible for the interpretations offered here. I am also grateful to P. V. Prabhu and S. Sinari of the National Co-operative Land Development Banks Federations in Bombay for their cheerful efforts of guiding all of us to the proper sources of information; to Dr. V. N. Deshpande of Sangli Town for acting as host and counsellor of students confronted with their first culture shock; and to Mohan Kulkarni who tirelessly acted as interpreter, local expert and big brother, and seemed to love it all.

I am further indebted to Franz von Benda-Beckmann for his encouragement to my students and myself to record our experiences of the multitudinous variety of actors and arrangements in the informal financial markets of developing economies. Without his stimulation and often flattering comments, this report would probably never have been written.

Dale Adams, Otto Hospes and René Bastiaansen commented on earlier drafts. Peggy Grossman edited the first draft of most of the manuscript while she was in Wageningen to study European agrarian legislation; she continued the ungrateful job of correcting my English grammar after her return to the University of Illinois. Janice Jiggins took upon herself to correct the remaining chapters of the book and suggested its title. Elly Mast did the typing and remained remarkably cheerful even when the umpteenth version arrived on her desk. I am grateful to them all.

I owe, however, special thanks to John D. Von Pischke who not only gave his continuous support, expert advice and comment, but made such a drastic and artful job of editing the second draft, that an almost entirely new report seemed to take shape under his hands.

F. J. A. BOUMAN
Wageningen, July 1988

Introduction

Financial institutions can play an important role in the growth and development of an economy by channeling funds from surplus sectors (savers) to deficit sectors (investors). Financial institutions will perform this role all the better when they constantly adapt their policies and instruments to the ever changing preferences of savers and investors and the challenges of new directions in the economy and the social fabric of society. Such challenges may stem from public as well as private initiatives. From differences between countries in the performances of rural finance markets, important lessons can be learned on the relation between finance and development. Because of Asia's exceptional economic record over the past few decades, Adams (1985:1) has suggested a study of Asia's rural financial markets.

Such studies are scarce and almost always limited to the performance of formal financial institutions such as banks, cooperatives and other development agencies that are officially monitored and regulated by the government and the central bank. The activities of financial intermediaries in the informal sector are seldom discussed. Development planners and economists share a belief in the superiority of formal institutions over informal intermediaries in playing a key role in the development of the economy. Such a bias is not limited to the industrialized countries of the Western hemisphere but extends to the Third World.

Third World countries are particularly committed to rural development. They devote time and energy to the creation of a rural network of formal financial institutions for channeling funds efficiently to sectors and target groups in the rural economy, in accordance with the national development plans. Later, the performance of these institutions is evaluated, praised or criticized without taking account of informal financial agents. Yet, the same policymakers who devise the national plans, readily acknowledge that the majority of rural households is dependent on informal intermediaries for their daily financial needs.

However much a government may deplore this dependency and commit itself to rectifying the situation, it is not realistic to ignore informal intermediaries nor to stake future rural development entirely on a rapid introduction of formal financial institutions in the countryside. The extent and nature of the dependency of rural households on the informal sector should be explored, the potential of informal financial intermediaries in promoting rural development investigated, the possibility of coexistence and complementarity of formal and informal finance agents recognized. The informal sector, as the oldest centre of enterprise, preceding the arrival of formal institutions, is a natural and legitimate part of rural financial markets and it is better policy to acknowledge than to deplore the fact. A study of the relation between finance and development should, therefore, encompass elements of both sectors of rural financial markets.

The district of Sangli in the state of Maharashtra seemed to provide the ingredients necessary for a case study of this nature. In semi-arid Sangli, public and private investment in irrigation, sugar factories and cane cultivation, in combination with a national sugar policy favourable to growers, have stimulated a local sugar boom ever since the late fifties. This boom has brought life to a hitherto stagnant agricultural economy. Since 1975, the dairy industry also has gathered momentum.

In an economy that is in rapid transition from a subsistence to a commercial orientation, the financial market requires extra attention. 'With increases in rural production, marketed output and specialization, a greater volume of financial intermediaries is required for efficient production and trade; poorly functioning financial markets could act as a serious break on development' (Von Pischke, 1983b:6).

India has always recognized the important role of the financial system in rural development. It has nationalized most commercial banks and urged them to play a more constructive role, while the Reserve Bank of India has provided financial assistance to rural co-operatives for decades. This has led to a strong growth of the formal financial institutions in India over the past 25 years. The same growth can also be noted in Sangli District.

During a short stopover in 1977 in Sangli Town, capital of the district, the author accidently learned of the increasing popularity of a fairly new form of local savings and credit association called *bishi*. This made him wonder about the effects of the sugar boom on the informal financial sector. According to conventional development concepts, formal instituions should have taken over part of the functions of informal intermediaries because the latter, it is argued, do not possess the resources and sophistication required by the expanded economy. Because of the successful introduction of sugarcane in a formerly dormant economy, more and more rural people and firms should begin to meet the eligibility standards of the banks. Concomittently, the size and format of informal financial intermediation should decrease proportionally. How, then, could the flourishing of a new informal model of savings collection and credit extention be explained?

The opportunity to address this question and study the relation between bishi and Sangli's development arose only in 1984 when graduates from the Wageningen University were able to participate in local research in India.

The initial intention was to study only bishis. Gradually, however, it became clear that other informal financial agents had also made a recent appearance. This called for closer observation of the relationship between the rapidly changing district economy and the performance of formal financial institutions at village and higher levels, and of how performance might be conditioned by India's rural development policies and financial system.

This manuscript, therefore, has become longer than originally planned. It contains not one but several accounts of operators in the informal finance market in Sangli District and discusses briefly India's rural finance policy and the district's economic awakening.

When the final draft neared completion, reports of two graduates in 1988 confirmed the continued popularity of bishis in the research area and the growing importance of milk collectors as lenders to cattle owners. They also reported the existence of ROSCA[1a] in the district. These ROSCAs, that had so far not been a subject of research, appeared to exist in three forms: the common type, in which the lot decides the rotation of the fund; an adapted lottery type with increasing dividenas paid to savers[1b]; and the auction type in which the fund is allotted to the highest bidder.

These 1988 reports from Jeske Kortenhorst and Kees Zevenbergen added new insights and re-affirmed, once again, the picture of a thriving and diversified informal finance market in the district. Unfortunately, the reports came too late for full absorption in this book and are only mentioned by way of an epilogue (see section 7.6).

The book is organized as follows. Chapter 1 briefly reviews the debate between proponents of formal and informal financial markets in Third World countries. Chapters 2 and 3 describe India's rural finance policy and the emergence of the sugar economy in Sangli. Chapter 4 gives a rough outline of the district's rural financial market. The subsequent sections review segments of that market: co-operative credit institutions in chapter 5, UCS in 6, bishi in 7, while an extensive account of pawnbroking activities is given in chapters 8 and 9. Chapter 10 describes the dairy boom and the lending operations of milk collectors. The final chapter attempts to summarize and evaluate the performance of these formal and informal financial agents against the background of the rapidly changing economic environment in the district.

Readers are reminded that the descriptions of India's rural finance policy and the development of a sugar and dairy economy in Sangli District, as well as the portrayal of the formal and informal financial agents are impressionistic and far from complete.

CHAPTER 1

Formal and informal finance, a review of the debate

This study investigates a number of formal and informal financial services in the context of rural development. Conventional wisdom is that formal financial institutions should replace informal ones to prevent economic stagnation and to expedite efforts of eradication of poverty. Lately, this wisdom has been challenged when discussing rural development of Third World countries (Cole and Parker; McLeod). Therefore, this study opens with a brief review of the current debate. The review mainly refers to the Asian situation, because the present survey is about financial institutions in an Asian country—India—and it is with regard to the Indian situation that the debate is most appropriate.

1.1. FINANCIAL MARKET COMPOSITION

Financial markets contain a formal and an informal sector. The formal sector is institutional and officially monitored and regulated, comprising the banks, co-operatives and special public credit institutions and projects. This sector is characterized by large scale operations, offering a wide range of financial services over a geographical expanse that exceeds regional, even national boundaries. It is protected by legislation, controlled by the central bank and supported by the state and the national and international banking community.

The informal financial sector is known by many other names: unregulated or unorganized market, informal finance, informal credit market, and parallel or indigenous financial market. These terms will be used in this chapter interchangeably.

It is difficult to give a clearcut definition of the informal sector because of its heterogenity. The term covers the activities of a great number of intermediaries such as the professional and non-professional moneylender, pawnshops, merchants and petty traders, landlords, shopkeepers, indigenous bankers and finance corporations; it also contains self help groups like guilds and other professional, recreational and religious organizations, burial associations and a great number of rotating and non-rotating savings and credit associations; finally, it covers private borrowing and lending arrangements between friends, neighbours and relatives. The essential characteristics of the informal market are its fragmentation and specialization, its localized and small scale operations which are beyond the reaches of official regulation and control by the central bank. Most of its transactions are not recorded in official statistics.

Despite this dearth of statistics and the general lack of information regarding composition and size of transactions, it is usually believed that the volume of rural informal finance in Asia exceeds that of formal finance, because of the widespread absence of banking institutions in the countryside.

1.2. FORMAL BIAS

The information gap on the activities of actors in the informal financial market, unfortunately, also excludes insight into the economic rationale of that market and the efficiency and impact of its operations. Discussions about informal lenders in Asia are usually charged with emotion and are seldom free from religious and ethnic prejudices. Asian literature often echoes the statements and views of politicians and bureaucrats, in which the informal sector is identified with the wrongdoings of the evil moneylender. Negative views about informal financial intermediaries, pushing ignorant and helpless farmers into debt in order to grab their lands, are widespread and difficult to dispel. Informal credit is conceived as exploitive and excessively dear; it is believed to encourage (conspicuous) consumption rather than productive investment that would benefit the economy; it is thought to provide only a limited volume and range of financial services. Most Asian policymakers are convinced of the moral and technical superiority of the formal financial sector over the informal one for supplying credit as an instrument of rural development.

Since the early sixties, development theories have put strong emphasis on rural development and self sufficiency of Third World countries in food production. Yet, and despite giant steps in improved technology, agricultural production has stagnated in a number of countries. This stagnation has been increasingly attributed to lack of capital. Agricultural credit has thus become an important tool to help resource-poor farmers improve crop yields through the application of modern techniques. To this was added a mounting concern for poverty alleviation and a more equitable distribution of the national welfare. Cheap credit is often advocated to improve the lot of the poor by developing small farms and other rural enterprises and to rectify the urban-rural imbalance. Programmes distributing cheap loans have become popular with donors and Third World countries alike. In Asia, given the prevailing attitude towards informal credit, such programmes were entrusted to co-operatives, banks and public sector agencies.

Experience with formal rural credit programmes has, however, largely been disappointing and has given rise to mounting frustration with the weaknesses displayed by formal financial institutions. These weaknesses are partly inherent in the institutional structure and culture and partly result from faulty interest rate policies. Distribution of cheap and easy loans of which repayment seemed unimportant, has become a popular theme with politicians to charm the electorate, further personal ambitions and realize ideological goals. But it failed to bring economic prosperity to the rural scene.

Cheap credit, by itself, does not create opportunities, it can't make an unviable enterprise profitable nor can it redress terms of trade between cities and the countryside. But it does bring a financial institution to an embarrassing position when its interest income is not allowed to cover its costs, and defaulters cannot be brought to repay their debts. The general result is the display of weaknesses that Von Pischke listed as follows: 'Limited access to rural customers, high cost of services, absence of savings facilities, financial nonviability, lack of active competition and inability to expand services to respond to and create opportunities' (Von Pischke, 1983a: 227).

1.3. NEW INTEREST IN INFORMAL FINANCE

The negative experience with formal rural credit programmes has sparked an interest in the operations of the informal financial market and its role in the mobilization and allocation of resources. Favourable comments on the workings of indigenous savings and credit groups as autonomous self help institutions (Nayar, 1973 and 1986; Bouman, 1979 and 1984) have brought home the fact that the informal sector consists of many other actors and modalities of financial intermediation than those of moneylenders, traders and landlords. There is a glimmer of recognition that even these and other rural powerholders, if the name befits them, can and do perform services that, far from being extortionate, may be beneficial to the rural economy and its participants.

Before the eighties, papers on informal lending were conspicuously absent from rural credit conferences—somehow it did not seem proper to provoke a discussion of such a controversial topic. Was this worthwhile when, indeed, the future belonged to institutional finance and the informal sector would soon become outdated? Today, the same conferences and workshops routinely devote part of their proceedings to a discussion of the merits and demerits of the informal sector. Representatives of developing nations now seldom express the view that to advocate support of ROSCAs is tantamount to efforts to keep their country from entering the era of modern finance—as was the experience of this author. Publications of Nayar (1982; 1986), Holst (1985), Timberg and Aiyar (1980), Chandavarkar (1985) and of scholars connected with the activities of the Ohio State University that started and promoted this debate (Von Pischke *et al.*, 1983; Adams *et al.*, 1984), have given fresh insights in the economics, if not ethics of informal lending.

While there can be little doubt of the formal sector's superiority over the informal one when it comes to financing large scale economic development and projects of national and regional importance, the role and strength of informal finance agents in small-scale rural economies and their subsequent importance to low income households should not be underestimated. In these penny economies the informal sector has a number of advantages over the formal one. It responds remarkably well to rural, particularly short term financing opportunities and allows low income people access to services not available to them elsewhere and at a relatively low cost. It can do so because the informal sector is the natural environment for rural people and antedates the introduction of formal institutions. People are, as it were, born into this sector, and this brings with it frequent face to face contacts, cultural affinity, and a great ability to adapt to the conditions of rural life. Unlike formal institutions, informal intermediaries do not need government subsidies to operate in a rural penny economy. They survive on the basis of competitiveness, financial viability and low cost operations.

New interest in the role of informal finance in the economy is demonstrated by relevant research by the Asian Development Bank in six Asian countries (Ghate, 1986). Further, the Asian and Pacific Regional Agricultural Credit Association (APRACA) has started a survey of Financial Self Help Organizations. This survey is part of a programme to promote linkages between banking institutions and such groups 'to improve the overall performance of rural financial markets in savings mobilization and credit delivery'. (APRACA News Digest: 2).

1.4. FORMAL AGENDAS FOR INFORMAL FINANCE

This new interest in the informal financial sector revolves, in principle, around the potential role of informal intermediaries in the mobilization of savings *on behalf of formal institutions*. Self help organizations in particular are thought to be an extremely useful conduit to channel a constant flow of rural savings to banks for economic recycling. Neither the recognition of certain advantages of informal over formal lenders, nor the negative experiences with formal rural credit programmes have apparently much shaken the conventional development wisdom that formal institutions are superior in allocating resources for the benefit of society. The limited geographical range of informal intermediaries is thought to prevent funds entrusted to them from flowing to the most economically advantageous uses. The report of the Indian Banking Commission of 1972 echoes the opinion that the growth of the economy requires more sophisticated institutions to satisfy the growing credit demands of the community. Holst, relying on economic history, does not expect a perpetuation of the current large share of the informal financial sector. 'In the course of economic development, formal institutions expand faster than informal institutions, taking over part of the latter's functions'. Drake even suggests that in the course of economic development informal financiers and loan associations become redundant (Drake: 140).

Belief in the superiority of the formal financial sector reverberates also in use of the term 'policy options' that are forwarded with regard to the informal sector. These options are widely believed to include: replacing informal credit agents by formal institutions through elimination (prohibition) or integration (institutionalization); benign neglect; support of the informal sector by improving its performance through regulation and competition with formal agents; promotion of linkages between the two, principally to channel funds from the informal to the formal sector.

The debate between proponents and adversaries of informal finance has barely started. Lack of information on informal operations and their rationale and impact have so far prevented a rational debate. Adams' earlier quoted suggestion that one must study Asia's rural financial markets to learn more about the relation between finance and development, is therefore in order.

CHAPTER 2

Rural finance in India

2.1. MAIN FEATURES OF FINANCIAL POLICY

India's financial and monetary policy has had a major effect on the development and operational efficiency of the financial infrastructure in the Sangli District. This paragraph indicates briefly the main features of that policy, thereby making use of a recent World Bank publication (Morris, 1985).

While the country's economy has become increasingly monetized, its financial system has come under tight control of the government and the Reserve Bank of India (RBI), which is the system's apex institution. The main objectives of this control are to achieve price stability through the regulation of the volume of credit and money in the economy, and to channel the flow of finance into desired social and economic directions in accordance with Plan targets (Morris: 55). One result of this policy is that India has been spared periods of high and persistent inflation; another is a lack of flexibility in the system to quickly adapt to changes in the economic climate.

The principal instruments of India's financial policy are increased public ownership and control of financial institutions, very detailed credit allocation policies, and a complex structure of lending and borrowing interest rates.

Both deposit and lending rates are subject to floors and ceilings based on a great variety of criteria (Morris: 65), and have considerable effect on the mobilization of savings and the direction of lending. The control of financial institutions ensures the effective allocation of credit flows according to detailed national and regional plans. Preferential rates of interest are set for priority sectors, regions and groups of borrowers. These include the government itself and what the government, in its efforts to eradicate poverty, has described as deprived social and economic sectors. Of these sectors, agriculture is the most important one.

2.2. RURAL FINANCE POLICY

Indian rural finance policy is characterized by a multi-agency approach, a proliferation of financial institutions controlled and supported by central and state governments, the RBI, and specialized national development agencies. Main suppliers of credit in the field are the co-operatives, which accounted for 59 per cent of total agricultural credit outstanding in mid-1980, followed by commercial banks and Regional Rural Banks supplying 39 per cent and 2 per cent respectively (Morris: 16). The different agencies and channels of short and long term loans are shown in Chart 1.

A second feature of rural finance policy are India's continued efforts to eliminate informal financial markets and harass its intermediaries through unfriendly legislation at the national as well as state levels.

Various Moneylenders' Acts, Debt Relief Acts and the Chit Fund Act bear this out. Indian authorities appear to discount completely the possibility that the informal financial sector may have a positive role to play in rural development and the continued well-being of small scale entrepreneurs.

It may be that they are reinforced in their opinion by the fact that researchers, scholars, planners, even private development agencies, persistently ignore the existence of informal financial markets beyond the mere admission that the majority of rural households in developing economies is still dependent on informal credit. Even a very recent discussion of India's financial system, published by an authoritative source like the World Bank, does not devote one single paragraph to the subject. Noting that 'systematic information on the size of the informal credit market is not available', the paper limits its discussion to what it calls 'the organized financial system' (Morris: 1).

2.3. RECORD OF FORMAL RURAL FINANCE IN BRIEF

Ever since the All India Rural Credit Survey of 1951 revealed the virtual absence of institutional credit in the rural scene, where

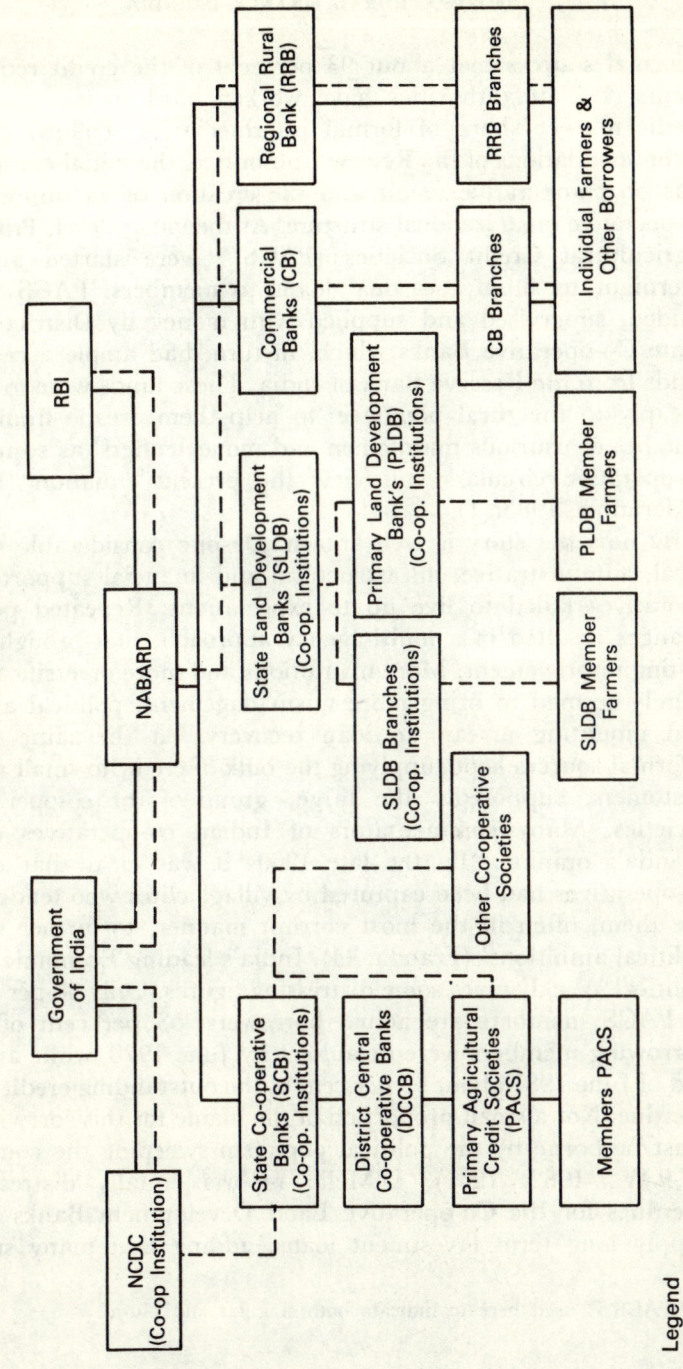

CHART 1
AGRICULTURAL FINANCE

Source: Morris: 17

Legend
— Short-Term Credit
--- Long-Term Credit

informal sources met about 93 per cent of the credit requirements, Indian authorities have worked hard to increase the credit market share of formal intermediaries. Following the recommendations of the Review Committee, the initial emphasis was on co-operative credit and the creation of an impressive co-operative organizational structure. At the village level, Primary Agricultural Credit Societies (PACS)* were started almost overnight to supply seasonal loans to members. PACS were guided, supervised and supplied with money by District and State Co-operative Banks which, in turn, had ample access to funds from the Reserve Bank of India. These funds were to flow cheaply to the rural populace 'to help them escape from the clutches of usurious middlemen and moneylenders' as so many co-operative circulars still view the present situation (LDB Federation, 1985: 1).

By now the story is well known. Despite considerable technical, administrative, infrastructural and financial support, co-operatives failed to live up to expectations. Repeated policy changes resulted in a multi-agency approach, that brought no lasting improvement. More institutions and more diversification merely seemed to bring more mismanagement, political abuse and mounting arrears in loan recovery. At the same time informal sources kept supplying the bulk of credit to small rural customers, supposedly the target group of the co-operative societies. Many commentators of Indian co-operatives echo Franda's opinion: 'By the late 1960s it was clear that most co-operatives had been captured by village elites who tended to use them, often in the most corrupt manner, to further their political ambitions' (Franda: 44). India's leading Economic and Political Weekly gives some distressing figures. Only 38 per cent of PACS members are actual borrowers; 53 per cent of the borrowing members were defaulters by June 1978, while at the end of June 1981 about 43 per cent of the outstanding credit was overdue. Not a small proportion of the blame for this sorry state must be borne by the political populism sweeping the country (E.P.W., 1982: 1261). D'Mello reports equally distressing overdues for the Co-operative Land Development Banks that supply long term investment loans, adding that many small

---

* PACS is used here to indicate both singular and plural.

farmers abstain from co-operative membership because they do not expect any benefit (D'Mello: 48).

The Indian government, alarmed by the extent of abuse and operational deficiencies, reacted by drastically pruning the co-operative sector, halving the number of village societies between 1961 and 1981. At the same time it introduced a change in institutional strategy. Commercial banking, which after successive nationalizations between 1955 and 1980 had become a virtual state monopoly, became instrumental in reshaping the rural financial structure.

In a ten-year time span, rapid expansion of the rural branch network resulted in a sixfold growth, from 2,200 branches in 1968 to 12,800 in 1977, increasing the percentage share of rural offices in the total from 25 to 43 per cent. In the same period a vigorous policy of resource mobilization boosted the total flow of deposits fivefold through the adoption of a number of schemes that favoured fixed and long-term time deposits. In 1985, over 28,000 rural offices dotted the landscape, constituting 57.5 per cent of total branch network and achieving a coverage of one bank office for an average of 16,300 population in rural and semi-urban areas (Reserve Bank, 1985a: 58).

Commercial Banks (CB) also rapidly expanded their rural lending. Of the 8 million borrowers in 1978, 3.5 million had agricultural loans (Ibid.: 60). CBs lend directly to farmers and rural enterprises, and indirectly by participating in the share capital of Regional Rural Banks (RRB) and by lending via the Integrated Rural Development Program (IRDP).

The principal task of a RRB is to lend to economically weaker groups that have no access to loans of CB. Started in 1975, the RRB network expanded rapidly to 8,200 branches in 1984.

The IRDP is set up specifically to alleviate rural poverty and focus on all sectors of the rural economy. In each district annual plans are drawn to provide selected beneficiaries with productive investment loans that contain a subsidy element of 25 to 50 per cent. All rural credit institutions participate in IRDP, but the bulk of the funds comes from CBs. The programme has a very high priority in India. Faced with immense political pressure to reach IRDP targets, CBs have gradually lost their grip on quality and supervision of lending. The banks, now charged with development functions at grassroots level, increased the share of priority sector loans from 14 to 27 per cent of total bank credit between 1968 and 1978; in 1985 this share had grown to 41 per cent. (Ib: 60). In 1985, 95 per cent of medium and long-term lending in agriculture was shared among CB, RRB and Cooperative Land Development Banks, with the Reserve Bank of India (RBI) providing refinancing through its executing agency, the National Bank for Agriculture and Rural Development (NABARD).

2.4. NAGGING PROBLEMS AND WEAKNESSES

Although these figures are impressive, there are many critics who find flaws in the performance of CB branches operating in rural areas, claiming that changes are more cosmetic than real and conceal much window-dressing (Shetty: 1419). The mobilization of financial savings has faltered because of low returns to depositors. This makes savers either turn to deposits with private companies or to the informal financial market that offers much higher returns; or deploy one's savings in assets such as gold and other commodities to escape the effects of inflation. 'The contribution of small savers to the savings pool could grow to significant amounts, considering their large numbers, if they are adequately compensated for their savings which, in the absence of social security arrangements, they surely need to accumulate over their working life.' (Reserve Bank, 1985b: 174).

Further, most of the resources of the institutional financial system end up in the public sector to satisfy the demands of bureaucrats, politicians and planners. This has kept CBs from meeting the growing credit requirements of consumers, distributive trade and small scale agricultural and industrial enterprise (Kumarasundaram: 795). Even the authorative Review Committee, that is usually mild in its judgment, lectures the financial policymakers, in 1985 still, that 'a beginning should be made in removing the misconception that trade is a low priority sector for bank finance as compared to industry in the present stage of economic development of the country' (Reserve Bank, 1985b: 307).

Again others complain that the amount of rural credit has not kept pace with rural deposits, which means that capital is, in fact, being diverted from backward regions. In 1978 less than one per cent of the under-two-hectare farmers of India obtained their loans from CBs (Franda: 69–70).

CB loan collection performance is bad, too. Like the co-operatives, rural banks have trouble recovering their money; re-scheduling of loans is common (Datta: 1361). In 1982 only 52 per cent of amounts due were recovered (Reserve Bank, 1985b: 72). Faced with heavy political pressure to achieve mandatory credit distribution targets, the financial system is blamed for too many subsidies, low profitability, a high degree of risk absorption and oversized bureaucracies that have to cope with outdated procedures and banking technology. All these have led to a mushrooming of unviable formal rural finance institutions (Kumarasundaram: 793; Morris: VII, 9).

In 1982 about 30 to 35 per cent of rural credit was supplied by institutional agencies, including the co-operative sector (E.P.W. 1982: 1261, note 15). This is certainly a long way from the mere 7 per cent of 1951. But it also implies that the rural populace is still largely dependent on informal lenders. Many Indians deplore this as unsatisfactory and critics blame the banking community, 'conditioned by their conventional moorings' (Shetty: 1248) for not having introduced any real innovations regarding the creation of a viable rural finance structure, despite all the rhetoric.

2.5. SUMMARY

India's rural financial system is characterized by a proliferation of financial institutions under tight public control. This reflects the government's attempts to regulate financial markets and direct the flow of finance into desired directions, in conformity with targeted assistance to deprived social and economic sectors, groups and regions.

Financial institutions are not allowed to make their own lending decisions nor to determine their own interest rates. These central interventions in financial market policy and credit allocation have reduced the operational autonomy and efficiency of banks, causing low banking staff moral, delays in administration, poor loan collection performance and low profitability. They have reduced the flexibility of the financial system to respond to changing economic conditions. The complex interest rate structure, moreover, appears to favor certain borrowers without necessarily contributing to promotion of priority sectors as a whole (Morris: 69). The setting of preferential interest rates for lending may even run counter to intentions by providing rich rather than poor farmers access to the cheap loans, as is so convincingly argued by a modern school of thought (Adams *et al.*, 1984).

The Indian institutional finance structure seems to exhibit very much the same weaknesses as those reported by Von Pischke (see Sec. 1.2, page 7). Judging from both the evidence and commentaries, supplemented by the critical observations of subsequent official Review Committees of the banking system, two major issues appear. One is the design of an appropriate institutional set-up for rural financial services. The other is proper appraisal of the role of formal financial intermediaries in this design *vis-à-vis* that of the informal sector. Critical observers of the Indian scene, moved by concern for the welfare of low income groups, doubt the willingness of the banking community to make an effort to lend to priority sectors, small cultivators and backward regions. 'To be made to function as an agent of change would require basic attitudinal changes among the bank personnel' (Shetty: 1446).

The question is whether this very proposition of banks as agents of change is realistic. The reluctance of banks to venture into risky tropical agriculture and small scale rural enterprise is not restricted to India, but is worldwide and well-founded. The fact that approximately 35 per cent of rural credit requirements in India, with its abundance of small agricultural units, is supplied by institutional agencies, could well be more of an achievement rather than a cause for dismay. The sometimes obsessive Indian concern with replacing the informal lender with a cheap and public-spirited institution seems to deny the underlying realities of a survival-oriented penny economy as described by Polly Hill for a region in Maharashtra. 'It makes no more sense to suppress (informal) rural credit-granting than rural buying and selling, each is essential to the economic health of the rural community'. Noting, too, the official opposition to 'village-generated credit which is all that is anyway available to poorer people', she remarks: 'Outsiders fail to comprehend the complexity of village credit systems or to realize that they can provide services which official agencies could never render. Partly as a consequence, they caricature the actors concerned as either villains or innocent victims—this being an attitude in particular tune with our times but which is also inherited from the colonial obsession with debt' (Hill: 216–7).

The research in Sangli confirms that the informal sector has its place in the spectrum of rural financial markets. Moreover, this sector shows considerable creativity in adapting its technology to the political climate and changing economic environment.

CHAPTER 3

The sugar boom in Maharashtra[2]

Much of Maharashtra State in West India consists of drought-prone areas with a largely non-commercial peasant economy, producing foodgrains for local consumption. Yields are low and farming conditions harsh. To improve these conditions and protect farmers from famine, the Bombay government began constructing canal irrigation on the Deccan Plateau late in the nineteenth century. Migrating Mali farmers were the first to grow sugar-cane, for which there was a ready market in India. Because no white sugar factories existed, the cane was manufactured into *gur*, the common type of artisan brown sugar, with rather primitive local crushing equipment in mobile plants, locally known as 'jaggeries'.

Cane production requires much water and labour and a large outlay of capital. It was therefore limited to farmers with access to all three production factors. With the gradual extension of canal irrigation, the arrival of higher yielding cane varieties and greater efficiency in water control, ploughing and harvesting operations, the growing of cane became an increasingly attractive proposition. In Maharashtra, due to soil conditions and the sunny climate in which cane thrives, yields of over 100 ton per hectare are recorded, with a sugar recovery rate from cane of 11 per cent, the highest in India.

Cane cultivation was further encouraged and came within the orbit of small scale production when Maharashtra State authorities facilitated and financed the establishment of co-operative sugar factories embracing both large and small producers. The first co-operative factory was organized in 1950. Assisted by an official policy of setting sugar prices low for the consumer and cane prices high for the producer, and with little competition from the private sector, these factories flourished. By 1982 there were 67 co-operative factories in Maharashtra, producing 32 per cent of India's total white sugar, 87 per cent of which comes from the co-operatives. The famous sugar boom was a fact, and it took Maharashtra less than 30 years. In Sangli, however, it arrived later and more slowly.

3.1. CULTIVATION OF SUGAR-CANE IN SANGLI

Sangli is a semi-arid district of 8563 square kms in Maharashtra State, (See Map 1). Four of its eight talukas (subdistricts) are officially recognized as drought-prone areas. Like other impoverished dry grain farming areas, it is, with 1.8 million people in 1981, only sparsely populated. Agriculture dominates the economy, a fact that is also reflected in the few major industries of the district, sugar, dairy and agricultural machinery. Sorghum and millet (60 pct) and pulses and groundnuts (20 pct) are the most important crops on a net sown area of approximately 600,000 ha. (Socio Economic Review). Because of the dry climate, yields are extremely low, less than 1 ton per ha. Regular, sometimes excessive drought conditions such as in the past ten years, have caused complete harvest failures, and much land is left fallow. As in similar regions in neighbouring Karnataka State, even large farms have difficulties producing a marketable surplus, and many small cultivators barely manage to survive (Hill; Schlesinger).

Sugar-cane cultivation came to Sangli relatively late. In Maharashtra four factors were instrumental in the sugar boom: irrigation facilities, a producer-friendly sugar policy with high cane prices, the establishment of (co-operative) sugar factories and a financial market for the provision of capital. Of these factors, availability of water is decisive. Sugar-cane is a thirsty crop and Sangli, unfortunately, suffers from an acute lack of water resources. Except for the Krishna and Warna rivers in the South-west, rivers are dry almost all year round.

Farm irrigation started here with water from private wells. In 1960, 37,000 of the 43,000 wet hectares were irrigated from wells, and in 1970 wells were still the most important source of irrigation. Up to then, surface irrigation was insignificant. The Koina dam in the Krishna river in nearby Satara district,

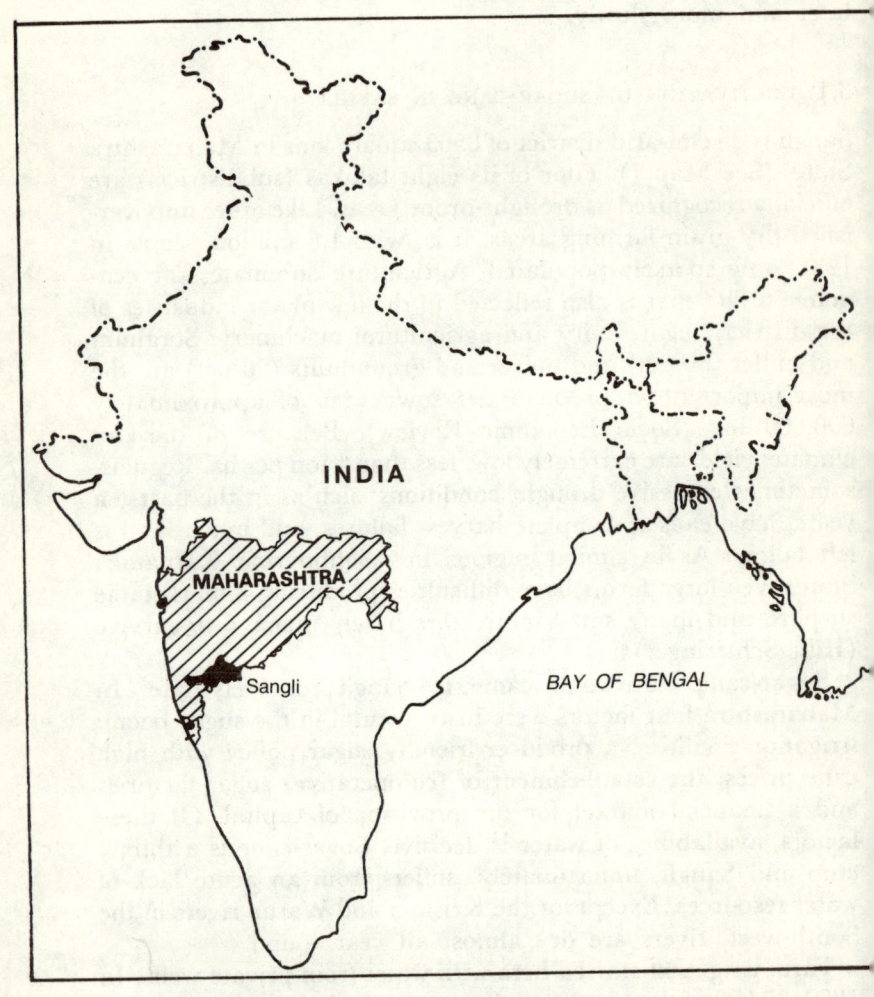

MAP 1. Location of Sangli District in Maharashtra, India

completed in 1956, was meant to provide irrigation from the freshly dug Krishna Canal. This canal, however, had only a potential command area of 6000 ha. Due to a low level of water in the Canal, actual use today is only 1000 ha. Another ambitious project is under way with the construction of a huge earthen dam across the Walwa river. When completed in 1988, it will provide Sangli with another 50,000 ha. of canal irrigation.[3]

Meanwhile, with the continuous failure of rains to replenish the groundwater table, wells had started to run dry in the 1970s. Sangli urgently needed other water sources for its parched farms. It found these sources in lift irrigation.

The principle of lift irrigation involves sucking up water from the river with high powered turbine motors and lifting it several meters to a pumphouse, from where it is transported through a series of pipelines to a distribution point in the field. From this point, water is delivered to the farms. Lift irrigation schemes in Sangli have command areas from 100 to 1600 ha. Several technical designs are possible, but all are capital intensive and beyond the financial means of individual farmers. In Sangli, therefore, they are financed either by the National Bank for Agriculture and Rural Development through the Co-operative Land Development Bank (LDB) or via a loan from the LDB to groups of farmers co-operating in so-called Lift Irrigation Societies. Supervision of the different operations is with the Irrigation Department of the Government, private engineering consultants or one of the sugar factories.

As more and more wells became dry, lift irrigation has become more important. The area under well irrigation has shrunk from 43,000 ha. in 1970 to 29,000 ha. in 1980, but the area under surface irrigation has increased from 15,000 to 31,000 ha. in the same period, keeping the net irrigated area constant at roughly 60,000 ha. or 10 pct of net sown area. The situation in 1985 had hardly changed. More lift irrigation schemes came into operation, and still more wells dried up. Of these 60,000 irrigated hectares, about half are under sugar-cane. These figures are estimates only; statistics vary according to the source and accuracy is difficult, particularly when so much land lies temporarily fallow. It is a fact, however, that sugar-cane has become the most important irrigated crop in Sangli, largely concentrated in a sugar belt between the Krishna and Walwa rivers in three talukas: Miraj, Tasgaon and Walwa (see maps 2 and 3). Here, one also finds the two major population centres: Sangli, the capital, and neighbouring Miraj city.

3.2. OPPORTUNITY AND RESPONSE

Irrigated sugar-cane can bring great opportunities, but in semi-arid Sangli it also requires entrepreneurship, imagination and a readiness to make the transition from seasonal dry foodgrain farming to perennial wet cane cultivation for a distant market. 'Before the rise of cane production, most villagers grew mainly foodgrains for local consumption, using family labour and the minimum of purchased inputs or borrowed capital. Cane cultivation requires the mobilization of hired labour and borrowed capital, the adoption of new methods and equipment, and the organization of new relations of production and distribution' (Attwood, 1984b: 39). While the government provided the initial facilities, the farmers had to respond by making the necessary changes.

Farmers in Sangli, large and small alike, took up the challenge willingly. Attwood and Baviskar, who have researched the origin, nature and impact of the sugar boom in Maharashtra since the early 1970s, attribute this to the historically loose and competitive stratification system of a sparsely populated semi-arid region, in which social mobility was common and enterprising entrepreneurship often rewarded (Attwood and Baviskar, 1984: 18–19). Irrigation and sugar brought new opportunities and changed the economic outlook of Sangli district.

Cane cultivation and processing are labour intensive and bring employment to many hands. First, there are the co-operative sugar factories. The oldest (1957) and one of the largest in Asia is in Sangli town and has a daily crushing capacity of 5,500 tons of cane. Two smaller ones date from 1970, and four others were established after 1983. Of the latter, only two have recently commenced operation on a small scale. Together, these factories employ about 3,500 labourers, roughly half on a permanent basis.

Harvest and transport of cane are organized and financed by the factories, with costs later deducted from payments to growers.

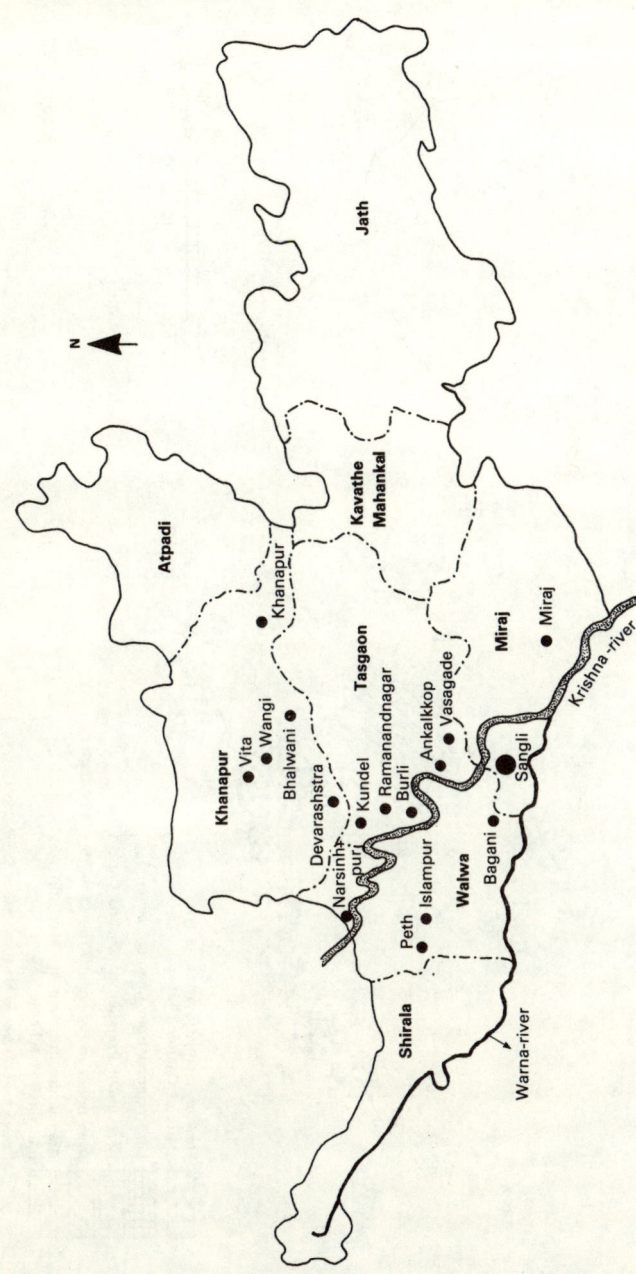

MAP 2. Sangli District, subdistricts and research villages

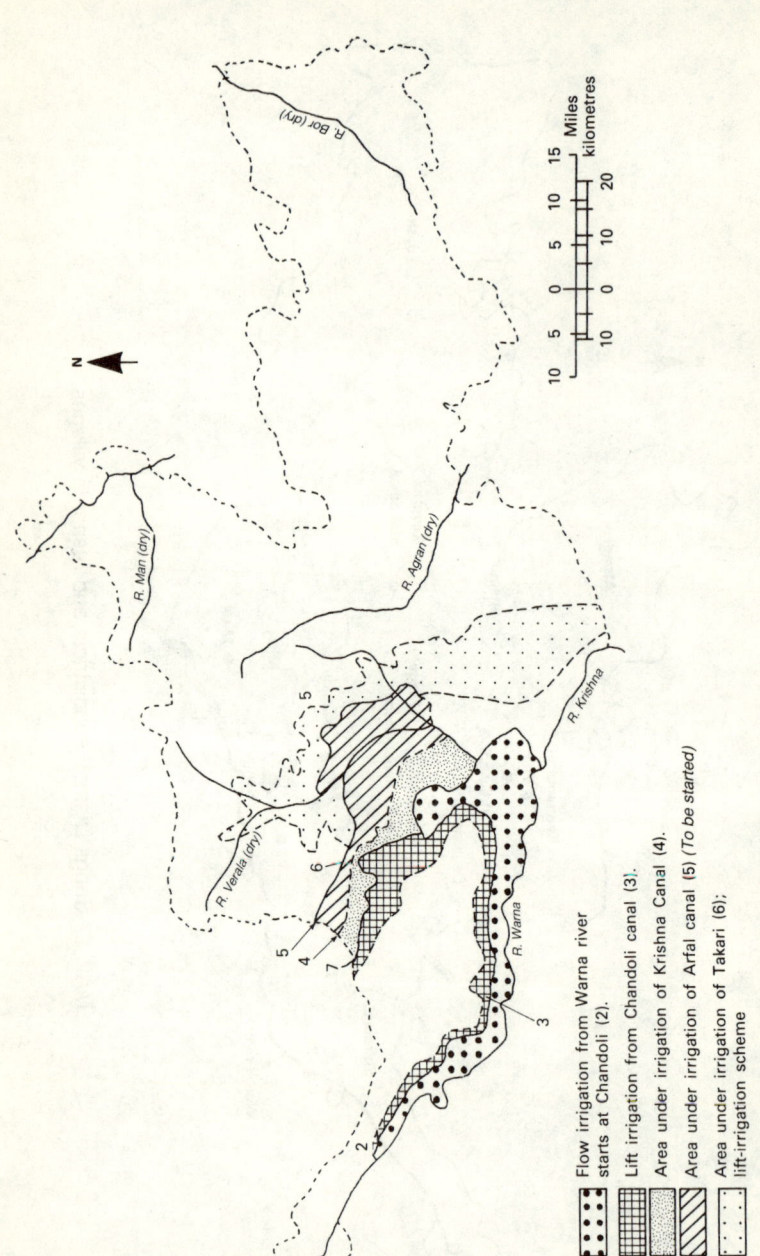

MAP 3. Irrigation projects in the district

These operations require about three workers per ton of crushing capacity (Rangaswamy) and hence some 30,000 additional workers keep the Sangli factories working in the six-month sugar season. Most of the labour force consists of migrants, but it also may include small farmers seeking additional employment as harvesters and oxcart owners (cane is transported not only by truck and tractor but also by ox-carts, which are a familiar sight in the landscape).

Spread throughout the sugarbelt are also many 'jaggeries', the traditional brown sugar plants. Sangli counted 241 licensed plants in 1985[4] and an unknown number of illegal ones operating without a license. Sugar extraction in these plants is crude, and about 20 per cent of the sucrose in cane is lost during the process (Attwood and Baviskar, 1984: 16). But the brown sugar fetches a good price in the market and the mobile plants operate cheaply because they can be moved close to the canefield, thus reducing transportation costs and losses in sugar recovery that occur when cane is not crushed immediately after cutting. Many farmers prefer the jaggeries to the factories because payment of cane is immediate; some hire the plants to process their own brown sugar and speculate in the market. These plants employ at least another 3000 seasonal labourers. Taken together, these employment figures are impressive for a district of only 300,000 households.

Both irrigation and the sugar industry have stimulated further development. Wells bring employment to well diggers and stone dressers. Canals, pumpsets, distribution points and tubes have to be maintained and repaired. Blacksmiths have to manufacture the iron ploughs and sickles, tillers and other implements[5]. Cane juice is sold in numerous small stalls along the road, and cane tops are fed to cattle. Molasses, the by-product of sugar processing, gives birth to alcohol distilleries. Molasses can also be turned into fodder, and dairy development in Sangli has blossomed, even in its non-irrigated dry zones.

Maharashtra State and Bombay City in particular, in a not too distant past, were dependent on Gujarat and the European Economic Community for milk and dairy products. This prompted the Government of Maharashtra to stimulate dairy development by importing crossbred cows since the early 1970s and Gujarat buffalos in the early 1980s. Medium-term credit for both types of cattle became available to farmers and priority groups from many financial institutions. Producers received a guaranteed and high price, while the consumers' price was subsidized. With a fivefold greater milk production than the local varieties, these cattle have caused a mini milk boom on top of the sugar boom. This happy development was doubly welcome because average yields of sugar-cane per hectare have declined seriously, because of increasing salination and exhaustion of the soil. How long the milk boom will last is uncertain. The market has been flooded with milk and since 1985 the district has had to destroy part of its surplus production (van den Bogaard: 10–20).

Co-operative sugar factories have become catalytic agents in the socio-economic transformation of the countryside, using their financial, technical and managerial resources to assist local people in new activities. Attwood and Baviskar cite examples of factory initiatives in lift irrigation, dairy plants, veterinary services, sugar research, infrastructure, banking, educational and health facilities, housing projects and stimulation of ancillary industries (Attwood and Baviskar, 1984: 54).

3.3. SUMMARY

Money and employment opportunities flowing from sugar-cane cultivation have given the economy of Sangli a healthy injection. Depending on drought conditions, the money flow varies from season to season. To give an impression of their magnitude, in the good 1981–2 crushing season 1.7 million tons of cane valued at Rs 436 million were crushed[6]. Add to this the money from the traditional jaggery plants and the salaries paid by the factories to their labourers, and the total is more than Rs 600 million. Income from milk sales of roughly Rs 400 million further increased the money flowing into the Sangli economy in that year. These money flows have brought higher standards of living and a quest for consumption alternatives and new investment opportunities. Because the sugar factories take care of harvesting and transport, cane farmers have spare time on their hands. Many take the opportunity to invest surplus money in new activities: high grade cattle, grapes, betel and rose gardens, small livestock (sheep, goats and poultry). Others invest in off-farm enterprise: shops, trade, transport, small industries, teahouses.

Rural households have also come to appreciate consumer durables including TV sets, refrigerators, sewing machines, bicycles and motopeds, giving life to garages, tailors and small repair shops. Undoubtedly, economic expansion will also have increased the demand for financial services. These financial services will be discussed in the next chapter.

CHAPTER 4

The rural financial market in Sangli District outlined

The Sangli economy is characterized by growing commercialization and financialization. Table 1 gives an impression of the extent of the financial market and the increased role of formal financial institutions in the district.

Table 1. Financial institutions in Sangli district, 1961–84.

Type of institution in the years	1984	1980	1971	1961
Commercial Banks	163			
District Central Co-operative Banks	98			
Co-operative Land Development Banks	13			
Urban Co-operative Banks	22			
Total Banks	296	242	102	45
PACS	518	518	521	511
Urban Credit Societies	133	91	10	6
Salary Earners Societies	92	78	NA	NA
Registered moneylenders	97	109	250	250
Bishi	NA	NA	NA	NA

SOURCE: Deputy Registrar of Co-operative Societies, Sangli.

At a first glance, financial developments in Sangli between 1961 and 1984 follow the pattern for India as a whole described in chapter 2. The number of PACS has remained stable over the years. This corresponds to the national trend of amalgamation of several unviable village societies into one single unit, while new societies were only started in hitherto unserved villages. By this measure the Indian Government hoped to stem the rising tide of unsatisfactory performance of co-operative societies, which was also evident in the district.

The number of banks increased sixfold, with the greatest growth between 1971 and 1984. This reflects the change in official Indian policy when, as a result of the disappointing performance of the co-operatives, the commercial banks became instrumental in reshaping the rural financial structure. In 1984 Sangli counted one bank for each one thousand households. Banks are concentrated (over 65 per cent) in the irrigated talukas in the sugarbelt.

Table 1 contains some surprises. The surprises are in the other segments of the financial market. The greatest growth, twentyfold, is recorded by the Urban Credit Socities (UCS) particularly after 1970 when the sugar boom had taken hold. A UCS is registered under the Co-operative Societies Act and must meet certain conditions set by the Commissioner for Co-operation. UCS* are subject to the regulations of the Reserve Bank of India concerning interest rates. Together with the other village society, the PACS, the UCS is generally considered to be part of the formal finance sector.

Yet there are reasons to set it apart as a semi-autonomous, semi-formal financial institution. The UCS is a much more spontaneous association than the PACS, which is practically a government organization. Because of the national policy that every Indian village should have its own PACS, the initiative to start one almost invariably originates with the official authorities. A UCS, however, is started by community leaders who do not form part of officialdom. Further, a UCS is a self help institution; it collects members' savings to make loans. Many PACS do not collect savings. Functioning mainly as a distributing agency of public credit funds, they are far from a self help institution. Being dependent on government largesse, a PACS has little autonomy with respect to loan policies. A UCS has more freedom to make its own rules and policy; control by the Registrar of Co-operative Societies is virtually non-existent.

The name Urban Credit Society is a misnomer. UCS are very much present in rural locations and handle savings as well as credit. They are also concentrated in the sugarbelt. Their rapid

---

* UCS is used here to denote singular as well as plural.

expansion since 1970 is even more remarkable than that of the banks. The latter is the result of official policy, the former is an autonomous movement and merits particular attention.

Table 1 also registers the existence of 78 Salary Earners Societies in 1980 and 92 in 1984. Earlier figures are lacking, but these societies reportedly are of rather recent origin. Because they operate largely in urban environments among civil servants and employees of sugar factories and the agro-industrial Kirloskar factory, they are not dealt with in this study.

The number of registered moneylenders, on the other hand, has decreased from 250 in 1961 to 97 in 1984. This is an interesting aspect in the development of the financial market of the district that merited a separate study. The results of that study are reported in chapter 9.

Not numerically represented in Table 1 are the bishis. A bishi is a people's savings and credit association that belongs to the informal sector. Because they are not registered, there are no official data of their numbers. However, their existence as a recent phenomenon was already recorded during a short visit to Sangli in 1977. Early in the present survey it became clear that this development had taken astonishing proportions, to the extent that bishis now outnumber both PACS and UCS combined. Study of bishi characteristics therefore became a major element in research into financial development in Sangli.

The research concentrated on bishi, UCS and PACS, which are all methods of delivering locally based financial services, each representing a part of the financial market. The PACS belong to the formal, the bishis to the informal sector, while the UCS, which is a semi-autonomous institution, comes somewhere in between. All three organizations operate at the grassroots level where procedures are less formal and information is more readily given than by banks, which have to preserve customers' confidence. Many bishis have the added advantage of regular weekly or monthly meetings that proved open to the curious outsider.

Yet, of the three institutions, the bishi is the least well known among academics and bureaucrats. No literature on bishis in Sangli could be found; indeed, scholars in Bombay and at the nearby University of Kolhapur doubted the very existence of a popular organization like the bishi. There is in India a widespread misconception of the nature of the informal rural credit market that forestalls intellectual curiosity. Informal finance is usually equated with the 'evil village moneylender', who should be replaced as soon as possible by a superior and honest public institution. Village-generated credit, however, is not only essential but also beneficial to the rural economy. It has many multi-coloured facets of which the bishi is both an important and revealing one that fully merits closer inspection.

PACS also were a major subject of study, being the formal financial village institution with the longest history. Popular experience with PACS as strictly an agricultural finance institution has probably hastened the birth and growth of UCS and bishi, when Sangli's stagnant agrarian economy turned into a more dynamic and diversified one. For a proper understanding of the role of PACS, India's co-operative credit structure is briefly discussed in the next chapter.

CHAPTER 5

Co-operative Credit

India has two co-operative credit structures, one for short and medium term lending and one for the long term lending. Short term co-operative agricultural credit flows through a three-tier structure with the State Co-operative Bank at the apex, the District Central Co-operative Bank (DCCB) as the middle rung, and the PACS as the final link at village level. PACS and DCCBs also distribute some medium term loans of up to five years duration. Money for these loans is collected by the State and District Banks—that compete with Commercial Banks for the funds of the public—and subsequently percolates down to the farmer via the PACS. Long term loans of five to fifteen years are provided by the Land Development Banks (LDB). The LDB structure is only a two-tier one, with a head office at State level and branches or 'primary banks' at local levels in each district. The apex State Co-operative Bank of Maharashtra is located in Bombay and has no offices in the district. Its main function is to act as a central bank for the other co-operative credit agencies. The ultimate source of refinancing of all co-operative credit is the Reserve Bank of India.

Because of the importance of the LDB as a source of finance for irrigation facilities in Sangli, its activities are briefly summarized.

5.1. THE LAND DEVELOPMENT BANK. (LDB)

The LDB has thirteen primary level offices in Sangli District in 1984; these do not accept deposits of the public. LDBs receive their funds by floating debentures in the capital market, while interim finance is supplied by the State Co-operative Bank, Commercial Banks and NABARD. Like the other co-operative credit agencies, LDBs raise share capital (to a limited extent) by withholding 5 to 10 per cent on loan amounts (LDB Federation, 1985: 7–8).

LDBs do not make seasonal loans but provide only long term credit for agricultural purposes according to directives and targets set out in annual development plans. Maharashtra State has historically received a lion's share of total LDB loans in India (LDB Federation, Five Year Plan 1985–1990, annex IV). Two-thirds of LDB lending in Maharashtra has been for investment in irrigation; followed by farm mechanization, dairy development, purchase of bullock carts and small livestock (sheep, goats, poultry, piggery; these loans are intended for the weaker sections of the community). In Sangli, the development of grape, rose and betel gardens is also stimulated. Loans to the weaker section of the rural community form about 55 per cent of total loan volume (LDB Federation, 1985: 4).

Ordinary borrowers pay a concessionary interest rate of 12.5 per cent but many, identified as preferential borrowers, pay less. LDB's lending margin is only 2 per cent; both the interest rates and lending margin are determined by the Reserve Bank.

LDB branches serve very large areas and have high operating costs and problems of viability. A report of the LDB Federation suggests a minimum outstanding loan volume of Rs 4 million per branch and a higher lending margin to function as a viable unit (Five Year Plan 1985–1990: 49–50). But even then the branches would operate as a money losing business, given the low quality of lending. Loan recovery performance of LDBs in general is bad, while that of Maharashtra State is among the worst in India, with e.g. a 35 per cent (!) recovery rate in 1980–1 and 51 per cent in 1982–3 (Ibid., annex VI). This bad performance is due, on the one hand, to the rigidity of a central loan planning process that frustrates the autonomy and flexibility of the individual LDB units with regard to loan decision making. On the other hand, the misuse of the co-operative credit system in India as a means of political patronage is widespread[7] and results in wilfull default on the part of big and influential borrowers. It reaffirms Adams' long standing critique of the use of subsidies in credit schemes (Adams, 1984).

Within Maharashtra State the repayment record of the LDBs in Sangli is somewhat better than average because of the marketing link with the sugar factories (LDB Federation, 1984: 64). LDB loans for lift irrigation schemes supervised by co-operative sugar factories, are recovered from the sugar-cane deliveries of the participating farmers. Table 2 illustrates LDB loan volume in Sangli District.

Table 2. LDB long term loan volume in Sangli District 1961–85 (in million rupees)

Year	Loan Volume	Year	Loan Volume
1961/62	2.4	1973/74	6.0
1962/63	5.0	1974/75	8.1
1963/64	5.7	1975/76	6.4
1964/65	8.1	1976/77	9.1
1965/66	8.4	1977/78	8.4
1966/67	10.6	1978/79	14.0
1967/68	16.9	1979/80	24.3
1968/69	32.0	1980/81	24.2
1969/70	35.0	1981/82	17.9
1970/71	17.4	1982/83	22.7
1971/72	14.6	1983/84	29.2
1972/73	7.6	1984/85	38.3

SOURCE: LDB District Branch Office, Sangli Town.

Volume of lending is low in 1961 and increases gradually until 1968; the emphasis is on well irrigation. In 1968 the LDB, with major assistance of the World Bank, starts to finance lift irrigation schemes, which results in a sharp increase in loan volume in 1968–70. This programme soon runs into difficulties and in 1971 lending subsides again. The extremely low figures of the next decade reflect the tightening of credit policy by the Reserve Bank under the threat of inflationary pressures. Business picks up again in 1980 with the increased participation of LDB in programmes to finance purchase of milk cattle and small livestock for disadvantaged groups in the rural community.

Thus the LDB, via supporting expansion of irrigation and the dairy industry, has been co-instrumental in promoting the sugar and dairy boom of the district.

5.2. DISTRICT CENTRAL CO-OPERATIVE BANK

The DCCB of Sangli District has its main office in Sangli town and operates 98 branches in the district. The principal task of the

DCCB is to supply PACS with money to make short term loans to farmers. These money supplies follow Annual Agricultural Development Plans, drawn up for each district, with sector-wise and scheme-wise allocations for each branch. Allocations are based on guidelines such as financial and manpower resources, institutional performance and targets that are set per crop and per scheme; advances from PACS to farmers also follow fixed scales per crop per acre. The Reserve Bank stipulates that at least 20 per cent of the bank's crop advances should go to smallholders with less than five acres; Sangli's DCCB managed to come to 30 per cent in 1984 (Brief note, 1985: 7).

Via the PACS, the DCCB also makes medium term loans of up to five years, sanctioned on the basis of minimum agricultural holding and repayment capacity. According to official guidelines, 50 per cent of these loans are to go to small farmers, but the fixed minimum acreage scales impede the realization of this policy. To a limited extent the DCCB also participates in various government subsidy schemes for disadvantaged groups.

The lending process thus follows a carefully outlined path of schemes and projects, with targets and beneficiaries that are decided at the central level. This plan-like approach to rural development tends to create a rigidity in the financial system that leaves little or no room for a dynamic response of individual branches to sudden, unexpected and unplanned for changes in loan demand.

The DCCB also finances co-operative marketing and processing societies, UCS and co-operative banks and the co-operative sugar factories. Loans to individuals are possible, too, like those based on the pledging of gold and jewellery. In Sangli the DCCB has gradually outgrown its main function as a PACS'-bank; the amount of 'other loans' in Table 3 has surpassed that of the short term crop loans. Comparison between loan volumes of this table and those of the PACS given in Table 4, also suggests that the money flows of the DCCB since 1975 increasingly bypass the PACS as loan distribution points. In 1982, for example, total loans volume of the DCCB was Rs 561 million, while the PACS distributed only Rs 146 million in that year. With the gradual growth of the sugar industry in the district, lending to the factories has taken increasingly important proportions.

Data in Table 3 reveal that lending levels were very low in

Table 3. Financial details of DCCB Sangli, 1955–84 (in million rupees)

Funds	'55	'60	'65	'70	'75	'81	'82	'83	'84
Share capital	0.3	2	7	10	13	21	24	27	30
Borrowings	0.4	7	21	33	66	13	87	147	65
Deposits	4	11	26	53	116	380	456	533	570
Loans									
Short term		12	31	59	91	103	141	143	174
Medium term	3	3	3	13	30	56	78	78	89
Others	–	–	5	96	33	84	211	340	262

SOURCE: DCCB, main office, Sangli Town.

1955, prior to the co-operative renaissance in India, but prior also to the birth of the sugar industry in the district. Lending explodes after 1965, as sugar-cane starts to dominate the agricultural economy. Clearly, co-operative credit has supported farmers' investments in the new crop.

Funds for lending come from three sources. Share capital, which includes the customary withholding of 5 per cent on loans; money borrowed from the State Co-operative Bank; and deposits of the public. These deposits are insured by the State and have become popular to the point of embrassing the DCCB with funds in excess of investment opportunities. Hence, it welcomes the demand of sugar factories to pre-finance their cane supplies.

Interest rates on deposits are prescribed by the Reserve Bank and in 1984 range from 0.25 per cent on current deposits to 11.5 per cent on fixed deposits of five years and over. Surplus money of the bank is kept with the State Co-operative Bank and receives a generous 11.5 per cent interest. DCCB lending rates vary considerably, from the concessionary 8 to 12.5 per cent for the various types of short and medium term agricultural loans to the commercial 17.5 per cent for non-agricultural borrowing.

The margin that the bank earns on its agricultural loans is small, 1.5 to 1.75 per cent. The margin for non-agricultural loans is higher. However, no lending margin can be sufficiently high to cover the overdues that the DCCB has to suffer. As has become habitual in the co-operative credit structure, these are so high that no bank would survive without the support of both the State and national governments. Overdues of medium term loans were

33 per cent in 1984 (Brief note, 1985: 5). Those of short term loans are discussed in the next paragraph.

5.3. PRIMARY AGRICULTURAL CREDIT SOCIETIES (PACS)

PACS are the officially recognized and supported farmers' lending institution at village level and the final link in the chain of short term co-operative credit between state, district and village. Credit in this chain flows from top to bottom according to the following interest rates structure: the State Co-operative Bank, Bombay charges 7–7.50 per cent on loans to the Sangli DCCB, which in turn charges 8–8.75 per cent on loans to PACS, which charge farmers 12–12.5 per cent. This 12 to 12.5 per cent compares favourably with the maximum interest rate of 17.5 per cent prescribed by the Reserve Bank in 1985. The difference of 5 per cent is to be regarded as a subsidy in co-operative lending to farmers[8].

Most PACS do not actively seek savings deposits and hence do not function as full financial intermediaries. More or less, they operate as retail agents of government loans for agriculturists. The PACS are very much a one way street that is hardly 'co-operative'. For this retailing service the PACS earn a commission of 3 to 4 per cent; slight variations are possible, depending on the type of loans and differential interest rates in schemes for certain groups of small farmers and other low income groups.

In Sangli roughly 70 per cent of PACS' short term credit used to go to the sugarbelt areas and was distributed among small and larger farmers alike. Since 1978 this percentage has dropped to 60. Loans for purposes other than sugar-cane are taken mostly by farmers with more than five acres. Up to 75 per cent is distributed in kind—most PACS, except those in dry areas, have their own fertilizer shop. PACS also provide some medium term loans for irrigation and farm equipment, bullock carts, dairy and small livestock development and gobar gas plants.

A PACS has a board of directors of at least seven members. Officially, this board is responsible for the society's policy and approves loan requests. In practice the secretary often is the omnipotent authority having the final say in loan decisions. Election of the chairman and appointment of the secretary have become very much a political affair. Most societies are affiliated to a political party and many a political career has started at the local PACS.

On paper, the time between request and approval of a loan is eight days, but in many societies it is much longer because the secretary or board are busy with personal and political affairs. The minimum salary of a PACS' secretary of Rs 400 a month is not conducive to diligent tending of the societies' business.

Table 4 records the loan volumes of PACS in Sangli since 1961.

Table 4. Short plus medium term loan volume advanced by PACS, 1965–82 (in million rupees)

Year	Loan volume	Year	Loan volume
1961/62*	16	1973/74	93.8
1963/64**	29	1974/75	100.7
1965/66	40.3	1975/76	59.3
1966/67	41.6	1976/77	65.6
1967/68	49.1	1977/78	70.3
1968/69	71.2	1978/79	73.8
1969/70	73.6	1979/80	86.8
1970/71	74.6	1980/81	139.5
1971/72	107.0	1981/82	146.5
1972/73	103.2		

SOURCE: District Statistical Abstracts, Sangli.
* Includes loans of other co-operative societies. Loans of PACS probably total less than Rs 10 million.
** Data for other years prior to 1965 are not available.

Total loans volume over the years is erratic. It was low in 1960 when the sugar industry was still in its infancy. It increased steadily from then until 1972, fluttered between 1972 and 1975, but declined sharply in 1976 and through the next five years, to pick up again in 1981. The principal reason for this decline is the mounting overdues that had reached 50 per cent in 1975 and forced the co-operative banks to severely limit their lending to PACS. In 1980 many co-operative debts in Maharashtra were waived, allowing to start lending again at higher levels than before.

Maharashtra co-operatives are notorious for their historically high default rates. Politicians like to attribute these to adverse weather conditions, but the minutes of the annual meeting of the State Co-operative Bank in 1976 voice another opinion: 'It is really intriguing to note that, on the one hand the State registers an all-time high agricultural production that outsteps the production targets and, at the same time, scarcity conditions are declared in some parts of 19 districts. I am particularly distressed by overdues in districts which raised cash crops with a perennial water supply' (Seven decades: 465). Franda, discussing the Indian experience in general, points to the numerous studies that show that the major defaulters are the wealthier and influential villagers 'viewing their overdue loans as a mark of social and political power and dare the co-operative society to take action', while in other cases 'local records have been blatantly altered to accommodate the errant lending habits of a village potentate' (Franda: 46). Examples of both were apparent also in the survey of 22 PACS and the 13 UCS in Sangli.

Senior co-operative officers also blame the sloppy performance of PACS' secretaries who have been described as 'a useless cadre' (Notes and News: 248). Unable to pay attractive salaries, many PACS do suffer from weak management that is incapable of appraising adequately and supervising a large volume of loans. Consequently, quality of lending is poor, loans disbursement delayed, and advances fail to satisfy farmers' needs. Poor management invites fraud and corruption, traces of which were found in 10 of the 22 surveyed PACS.

The co-option of the co-operative movement by politicians has had disastrous results in India. Franda quotes the dry prose of a Rural Credit Review Committee to drive this point home: 'The intrusion of politics into cooperatives has meant that individuals not belonging to particular parties or factions are virtually outside the scope of cooperative credit. For this reason, the assumption that cooperatives could meet the credit demands of all creditworthy cultivators, has proved untrue or only potentially true. Even where, technically, all could be served, politically only some have benefitted' (Franda: 47).

Such a sorry state of affairs undoubtedly creates a climate that is ultimately conducive to the introduction and acceptance of an alternative financial institution. In a sense, PACS in Sangli have paved the way for the popularity of bishi and UCS. Practically nowhere in India has the co-operative movement made such inroads in rural village life as in Maharashtra, which prides itself of a co-operative society in virtually every village (Progress: 2). The co-operative movement has become the training institute for ambitious politicians, actively discouraging collection of overdue loans. The artificial conversion of unpaid short term loans into a long term one is common practice in PACS. Maharashtra's politically inspired leniency towards co-operative mismanagement has already brought repeated conflicts between the State and the Reserve Bank. The latter appointed a study team on co-operative credit in Maharashtra in 1972. The State authorities subsequently refused to follow the advice to amalgamate and/or liquidate the many unviable village societies (Seven decades: 429). Instead, Maharashtra followed its own 'rehabilitation scheme for non-wilful defaulters', which amounted to forgiving bad debts and enabled PACS to continue making fresh loans (Ibid., 443). When co-operative debts were waived again in 1980, this decision was criticized severely in local newspapers (see Appendix 2).

Still, Sangli has a better collection performance for co-operative debts than some other districts within the State. While overall recoveries in Maharashtra consistently hovered between 50 and 60 per cent after 1980, Sangli district recorded levels between 60 and 70 per cent. This is attributed to the marketing link with the sugar factories that deduct loan advances from payments to cane cultivators. The figures of 1980–1 serve as an illustration. The recovery rate of PACS' debts in the district was 63 per cent against 68 per cent in the three subdistricts of the sugarbelt: Miraj, 69 per cent, Tasgaon 72 per cent but Walwa 62 per cent (District Statistical Abstracts: 73). The lower figure for Walwa is intriguing. Could it be that cane growers in Walwa avoid repayment by having their cane processed by the jaggery plants rather than by the factories? It is not uncommon that farmers have to wait almost a year for the final payment of their cane by the factory, while payment by the jaggeries is immediate.

The financial viability of PACS appears to be measured by its ability to collect loan interest, which is deducted from loan capital in advance. As long as interest is paid, overdues do not threaten a PACS' financial viability directly, because the societies only act as the distributing agent of outside money. The real burden of unpaid loans falls on the institutions higher in the co-operative structure. Indirectly, however, a PACS is affected. Members with unpaid debts are not eligible for new loans, therefore the number of borrowing members must decline steadily. In Maharashtra the figure fell below 30 per cent in 1981 (Brahme: 7); in the Sangli survey of twenty-two PACS in 1985 it had dropped to 25 per cent (Groot Kormelink: 7).

In addition to credit, PACS may offer marketing services, have a 'fair price' consumers' shop and sell fertilizers. These activities may bring additional income, but the 3–4 per cent commission on loans is vital for survival. Most PACS operate cheaply, usually from rent-free premises and with only a small staff. Salaries are so low—a few hundred rupees per month—that the societies can not attract qualified staff, which is why so many secretaries devise their own ways of supplementing salaries. Annual overhead costs vary between Rs 15,000 and 20,000. A 3 to 4 per cent loan commission thus implies that an annual loan volume between Rs 375,000 and 650,000 is needed. With the steadily declining number of eligible borrowers, there is hardly any PACS with such a loan turnover. Already in 1972 the State Co-operative Bank considered 60 per cent of the societies within the State non-viable on the basis of the same criterion. In June 1980 the number of non-viable societies in Maharashtra had risen to 74 per cent (Seven decades: 511). That the PACS continue to operate and even increase the lending flows (see Table 4), is only because they are kept alive artificially. The distribution of cheap co-operative credit to farmers has been a favourite theme of politicians in India for a long time. According to information that the author received from India early in 1989, the Maharashtra government has voted in 1988 to write off a large volume of co-operative and bank loans. The same decision has been taken before in the State. It destroys repayment discipline and strengthens borrowers' belief that defaulting on a loan from a public institution will eventually go unpunished. The very lenient Indian loan policy fosters a culture of evasive tactics and manipulation of loan officers and slowly erodes both lenders' and borrowers' moral behavioural code.

CHAPTER 6

The Urban Credit Societies*

A UCS (Urban Credit Society) is a semi-autonomous self-help institution for saving and borrowing. It is registered under the Co-operative Societies Act and comes under the jurisdiction of the Commissioner for Co-operation. Supervision by that office is, however, superficial. Every UCS is virtually free to devise its own policy and arrange its own affairs, as long as it keeps the interest rates on savings and loans within the limits prescribed by the RBI.

The first UCS in Sangli District was originally a reaction to the PACS of that time, which were lending institutions for farmers only. It was founded by tailors of Islampur, a small town in Walwa taluka, and registered as a co-operative in 1918. Tailors also founded the second UCS in nearby Peth in 1927. Already at that time caste and politics played a role: in both UCS Marathas were excluded from membership for fear of domination by this powerful caste of farmers.

Few UCS were registered thereafter; in 1973, societies in the district numbered only 13. The real explosion of UCS began in 1974, a phenomenon that clearly suggests a relationship with the expansion of the economy caused by the sugar boom. Between 1974 and 1984, 120 new societies were registered; 75 per cent of these are located in the sugarbelt. Beneficiaries of the sugar boom had money to spend and were looking for new production and consumption alternatives. When the existing financial institutions could not accommodate them, they started their own institutions as the tailors had done decades earlier.

One finds UCS registered in areas, despite the presene of numerous other financial institutions. A typical example is

---

* This chapter is based on data collected by Hospes. Data on UCS Ambika at Wangi come from Gerner.

Ankalkhop, population 10,000, with a DCCB, one Commercial and two Co-operative Banks, one PACS and three UCS; or Bagani, population 8000, with a DCCB, a Commercial and a Co-operative Bank, one PACS and one UCS (see map 2 for research villages).

Most informants, including farmers, emphasized the following as the main *raison d'être* of the UCS: the existing institutional system of PACS, LDB, DCCB and even some Commercial Banks, narrows its lending to agricultural activities, to comply with official development blueprints. Procedures are cumbersome and highly formalized, and loan applications take a long time to process. Both Commercial and Co-operative Banks favour large loans based on firm collateral, which is not readily available. Applications of borrowers with an overdue PACS loan are refused. Small loans, either to start or continue a new activity in distributive trade, craft or transport, or for education, consumption and ceremonial expenditure, are almost unavailable. In addition, PACS loans are often issued (too) late; and final payments of sugar factories to cane growers take a long time to arrive, forcing farmers to look elsewhere for interim credit. People, therefore, make use of different financial institutions interchangeably.

In the view of co-operative policymakers of Maharashtra, UCS are foremost a financial institution for non-agriculturists. 'Membership of a UCS would normally be drawn from persons engaged in different occupations such as trade, commerce, small scale industries, self-employment, employment in Government and semi-Government bodies'. They are intended to operate in urban or semi-urban centres. A UCS may be formed in an area with a minimum non-agricultural population of 7000, with an initial membership of 200 and a starting capital of Rs 12,000. These requirements are meant to safeguard the economic viability of the UCS by restricting overlap of operations and avoidance of dual membership and multiple borrowing[9].

Despite these well-intended goals, UCS in Sangli have been registered in centres of fewer than 7000 persons, with two or more operating in the same area. Membership of farmers, many of whom have business interests beyond their farm, is very common. How do these societies perform?

Several characteristics are essential for viability. These include low operational costs, growing membership and funds, sound lending and recovery performance, together with fair and equal treatment of all.

Membership in most of the thirteen surveyed UCS ranged from 200–500 persons (mostly males) of all professions. Although a UCS is not meant for farmers, they formed the majority and even dominated the board in nearly half of the societies. Only three societies had a membership of over 500, increasing steadily over the years. Such steady increase is vital for growth, but it also harbours the danger of faction. 'The larger the organization, the more likely it is that cleavages will emerge and manifest themselves in competitive claims for control of the organization and its benefits' (Esman and Uphof: 197). This can only be countered by a strong and trusted leadership, supported by a professional staff averse to favouritism. As will be discussed later, this was seldom the case in the UCS of our survey.

Share capital and savings deposits are crucial. Without sufficient funds, loan demands cannot be met. Share capital amounted to over Rs 100,000 in three UCS, but stood below Rs 50,000 in all others. The three with the most capital were also the ones with the largest membership. One of them relies solely on share capital for funds and has no other savings programme. This UCS of better-off sugar-cane farmers in fully-irrigated Ankalkhop, Tasgaon taluka, aspires to convert into a Co-operative Bank to broaden its financial basis. A share capital of Rs 250,000 is required to establish such banks and the society had already collected Rs 225,000 by June 1984. The other two large societies had each collected over Rs 100,000 in savings deposits. One of them is Ambika in Wangi, Khanapur taluka (fairly dry country); this UCS is reputedly the model of the whole district. In addition to people of Wangi, it serves seven neighbouring villages and draws deposits from over 1000 non-members, a sure sign of faith in the trustworthiness of Ambika. This UCS, too, wants to expand into a Co-operative Bank, and registration was imminent in 1985. The other large UCS is located in Ramanandnagar, Tasgaon, a thriving agro-industrial community of 8000 people in the sugarbelt, with the Kirloskar agricultural machinery factory and many traders. It, too, offers savings facilities to non-members.

A distinctive feature of some UCS in Sangli is the 'pigmy

deposit' scheme, in which tiny sums of money are collected daily or weekly from savers by a commission agent. This provides an UCS with a constant input of funds, however small, and also gives members easy and cheap access to a depository for safekeeping. One should bear in mind that rural people have a savings need as well as a credit need (Vogel, 1984). The agent is paid a commission by the UCS of 2.5 to 3 per cent on collected funds, sufficient to motivate him to maintain and enlarge the field of operation. The UCS that performed best in the survey, Ambika in Wangi, kept three agents busy, together collecting Rs 40,000–50,000 a month. One of the worst UCS, in contrast, had to fire its agent because of fraud.

The daily deposits usually pay between 6 and 7.5 per cent annual interest to savers. Beyond this scheme, however, a few UCS offer a diversified savings programme, with term deposits ranging from one month to five years and longer, and with annual interest rates climbing correspondingly to 12 per cent. In this way they try to meet the preferences of all categories of savers. Prominent in this respect is Ambika Society, with a healthy balance of over Rs 1 million in savings.

Savers can be lured by attractive dividends, and a few UCS do compete with banks by offering 0.5 to 1 per cent higher interest on deposits. But the savers' prime concern is security; here the banks have the definite advantage that their savings deposits are insured by the State, while those of UCS are not. With the exception of the 'big three' in our survey, all other societies had no significant savings programmes. This was either because such programmes had collapsed (six cases) or because the societies were only registered since 1984 (four cases) and had still to develop them.

A society with insufficient funds may still borrow from a DCCB, but only to the limit of its own share capital. Such borrowing will cost 15–15.5 per cent interest, that is 7 per cent more than a PACS has to pay for DCCB credit. This illustrates the preferential treatment by the Indian government of farmers in general. In interviews, members demonstrated their own logic in explaining this difference: 'Surely the PACS charge a lower interest rate than our UCS, but that is because farming is risky'.

Members, in turn, can borrow from their UCS at 18–18.5 per cent, equal to the rate they would have to pay for a Commercial

Bank loan, if available[10]. Because of the 10 per cent 'witholding' rule, however, borrowers do not receive the loan in full but have to turn 10 per cent of the loan amount into the society's share capital. This raises the effective loan interest rate slightly.

Most lending is limited to one year only and most UCS prescribe maximum loan amounts between Rs 500 and 2000. Every loan requires two guarantors, both fellow members of the UCS. The better-performing societies charge lower interest rates (16–17 per cent), offer higher loan maximums (up to Rs 5000), and a longer repayment period (up to three years). Our model UCS in Wangi does a thriving lending business based on gold as collateral. Loans are for all kinds of purposes in the spheres of production, consumption, and ceremonial expenditure in religion and marriage. Farmers also use these loans as interim credit, as explained earlier.

Loan default is the biggest problem plaguing the societies. Like Franda did elsewhere (Franda: 46), Hospes reported that Maratha farmers—the dominant caste in Sangli—took a good deal of pride in overdue loans, which they viewed as a mark of power and influence (Hospes: 64). One informant described the general attitude within his UCS eloquently: 'Members reasoned that, since there was no need to repay PACS loans, why should UCS-debts be repaid?' As for our three large societies, the repayment record of two of them seemed impeccable. The third one listed court expenditure in its books, implying defaulters, but still showed a substantial profit. One society looked dubious, with its office closed most of the time, while the collection agent had defrauded to the extent of Rs 12,000. Five societies were moribund; four others were still too young for any realistic appraisal.

Sad experience with UCS has already prompted the DCCB to exercise prudence in lending to them, as Table 5 illustrates.

Despite the increasing number of UCS throughout the period, fewer have been refinanced with less and less money. The contrast with the PACS (last column), where loan default is as problematic as among UCS, is striking, showing once again India's preferential treatment of agriculturists. While the PACS receive more and more, the UCS receive less and less credit, although the DCCB can lend to an UCS at 15 per cent but to a PACS at only 8 per cent.

Table 5 Outstanding DCCB loans to UCS and PACS in Sangli District 1980–84 (in million rupees)

Date	Urban Credit societies			PACS*
	No. of UCS	No. of borrowing UCS	Loans outstanding	Loans outstanding
30.6.80	91	60	1.48	86.5
30.6.81	102	60	1.63	102.9
30.6.82	117	42	1.25	141.2
30.6.83	119	38	0.86	142.5
30.6.84	133	47	0.95	171.3

* The number of PACS is about 520 throughout this period 1980–84.
SOURCE: Hospes, using annual reports DCCB Sangli-town (Hospes: 56).

The older UCS in particular, established in the 1970s, have performed badly and collapsed. As with the PACS, the apparent cause seems to be petty politics. Rival local powerholders set up and sometimes largely finance their own institution to gain support, and follow this up with loan favours. Significantly, collapse of UCS was prominent in towns with more than one society, each catering to its own political faction. From Hospes' survey the following statements are quoted: (Hospes: 74 a.o.).

'I will never join this society because it has the wrong political colour' (a shopkeeper).

'We lend on the basis of repayment capacity, unlike the PACS that lend on the basis of relations and favours' (Chairman of a well-established UCS).

'Politically active farmers are refused membership' (Secretary of a bishi, an informal financial organization, see next chapter).

'A politicized UCS is not a good thing because it results in bad repayment. Directors cannot exert pressure on defaulters for fear of losing votes' (a grocer).

'I use this UCS to promote Congress (I) Party and have better relations with Party followers'. This frank statement comes from a UCS Director in the politically active town of Kundal; the rival UCS in the same town championed the cause of the Peasants and Workers Party. Both societies are now moribund.

Other societies have been started specifically to counterbalance the dominance of a particular caste group, notably the Marathas who represent the strongest group in Sangli in numbers as well as in economic power. The former manager of the highly reputed Ambika Society in Wangi admitted that his UCS was created to counter the discrimination exercised within the local PACS controlled by two rivaling Maratha groups. No favours were allowed in his society, particularly not to Marathas, and this seemed to contribute significantly to Ambika's success. Others had similar stories to tell, reminiscent of the experience of the tailors, establishing the first society in Islampur in 1918.

Like the PACS, UCS operate cheaply from practically rent-free and sparsely furnished premises with a small staff earning low salaries. Annual overhead costs vary between Rs 5000 and 12,000 depending on the age of the UCS. New ones start with a small membership, little capital, and only one secretarial position. Overhead has to be met from the spread between borrowing and lending rates of interest, which averages 6 to 7 per cent[11]. This works out to a minimum annual loan volume of approximately Rs 80,000–200,000 necessary for financial viability. A loan default as little as 5 per cent, which reduces the spread accordingly, is already disastrous for survival.

Survey impressions are summarized below.

1. In theory, a UCS is an association of non-agriculturists. In practice, many farmers have become members and may even dominate a society's affairs. The cane farmers were the ones who benefited mostly from the sugar boom and had money to spare for consumption and investment in a side business, and needed a financial institution to accommodate them. Unfortunately like the PACS, the UCS has also fallen victim to the aspirations of ambitious politicians. Politics are the main reason that societies abandon sound business principles and perform badly. In such UCS the savings function is neglected and loans are granted in exchange for votes. The three UCS with the best performance had a sound savings programme and were able to sanction larger loans for longer terms than the other societies.

   The easy-loan routine of most UCS corrodes the already low repayment discipline of borrowers who have come to accept credit favours as their due. For this reason alone, the suggestive name Urban *Credit* Society, that leaves out the necessary savings dimension, is an unfortunate choice. The disfunctional role of politics was most evident in the older societies, especially in town where two or more UCS were founded almost simultaneously by political rivals.

2. During the survey, however, the impression was gained that people had come to realize the eroding influence of politics on financial stability and performance. Some of the newer societies have been championed and are directed, not by politicians, but by motivated professionals with experience in managing a financial institution. In this motivation, one seemed to detect traces of Hill's 'colonial obsession with debt' (Hill: 216), a compassion for fellow countrymen who earn too little but spend too much, and whose burden could be relieved by a properly functioning savings and credit society. Whether or not such belief is justified, professionalism in financial management is a better basis than politics for a successful UCS. Even in towns where UCS had collapsed earlier, there are signs of revival, combined with innovation. For instance, in places where a collection agent had committed fraud, the new agent had to deposit a guarantee of several thousand rupees from which future shortages could eventually be met. New savings pass-books and regular controls were other safeguards against possible deceit.

3. Many times, relations between UCS and bishi, in the form of dual membership or otherwise, were encountered. Two of the thirteen UCS had formerly been bishis that had evidently outgrown the needs of their membership. In other situations, board members of the UCS had been or still were chairman or secretary of a local bishi. During a short visit to Sangli town in 1977, the author came across a UCS whose promotors still continued their bishi business as a sub-group. Savings and lending technologies of bishi are copied in UCS and vice versa, apparently in an effort to find a proper balance of organizational viability.

CHAPTER 7

The bishi, a self-help organization of the informal finance sector

7.1 NATURE AND ORIGIN OF BISHI: OF ROSCA AND RESCA.

Bishi is short for bishi mandal, the Marathi vernacular for money club. It is the collective name for the rotating, as well as the non-rotating, savings and credit association. In these associations, members make periodic contributions that are pooled in a fund from which loans are made. In the rotating bishi, the total fund is given to each of the members in rotation until everyone has had a turn; hence, only one loan is made each time the members convene. In the non-rotating bishi several loans are made from the pool of savings. Over time, some members may take more than one loan; others may never need to borrow from the fund and use their society merely as a savings club. The non-rotating bishi resembles the Credit Union of the Western hemisphere.

Both types of societies are common in developing countries, but it is the rotating savings and credit association that has captured the imagination of researchers and policymakers worldwide. Internationally it has become known by its acronym ROSCA (Bouman, 1979). No acronym yet exists for the other type of society that is just as (or even more) universally popular as the ROSCA. To distinguish between the two, it may be useful to introduce the term RESCA, for Regular Savings and Credit Association.

A favourite ROSCA pattern is the one of twelve participants making monthly contributions, which will take exactly one year for completion. If the individual contributions amount to $ 10, one of the participants will pocket a fund of $ 120 each month.

ROSCA with weekly or even daily contributions—and therefore weekly or daily drawings—usually have more members, but the cycle seldom exceeds one year.

The capital of the ROSCA-fund does not grow during its lifecycle*. Each time the members convene and submit their contributions, a new fund is formed but is depleted immediately again. The RESCA-fund, on the other hand, grows over time, while loans are taken and repaid at regular intervals.

Only at the end of the lifetime of the RESCA—usually one year, too—is the accumulated fund distributed among all members. In the ROSCA there is nothing to distribute at the end of the cycle. At the final meeting the last member in the cycle takes the total fund, which then represents his accumulated savings over the ROSCA's lifetime. The fine point of the ROSCA is that all members, except the last one, receive their accumulated savings ahead of time. It is probably this peculiar trait that explains its appeal to both participants and observers.

The Indian ROSCA is commonly known as *chit fund* or *chitty* and is described most extensively by Nayar (1973, 1986). Its origin antedates the establishment of modern banks. Originally, contributions were in kind, paddy or rice. This made the chitty very popular among women, who saved a handful of rice from each daily meal to contribute to their rice bank. They invested the proceeds of the chitty in gold ornaments, household utensils or small livestock (Nayar, 1986: 32). In Sangli, this type of ROSCA became known as *grainbishi*, because contributions were in grain. It still functions in a few places in the district and the state. Grain banks also exist elsewhere in the Third World as a place to store the local staple and to lend it out in times of distress; but these are normally not of the rotating type.

With the monetization of the economy, contributions in cash gradually replaced those in rice or grain. The chitty became popular among professions other than agriculturists, and its nature changed from a pure savings club to a savings and loan society (Nayar, 1986: 34). The Indian money-ROSCA knows three basic variations of the rotation principle. The first is the lottery type, in which lots are drawn to decide whose turn it is to

---

* In some countries, however, members may agree to increase contributions—and hence the fund—to match inflation.

receive the fund. The second is the lottery with discount, which works the same way, except that a small sum is deducted from the fund and distributed among members who have not yet received the fund. The discount can be seen as a form of interest on savings and is an improvement on the simple lottery system that gives members no such dividend. This is annoying in times of inflation and especially to the players who receive the fund near the end of the cycle. In the discount-lottery, the dividend to savers increases towards the end. An example of this type of ROSCA is given below. The third type is the auction-ROSCA in which members bid for the fund. Eventually, the fund goes to the bidder offering the highest discount, which is subsequently divided among the other members. Combinations of both auction and lottery may occur, too.

Practically all ROSCAs encountered in the district were of the first two variations, the lottery with and without discount. Some people admitted familiarity with the auction principle, but claimed to dislike it because it introduces an unacceptable measure of competition between members. The lottery-ROSCA is still alive in Sangli but mainly among women, perhaps because of its earlier link with the ancient grain bank. In the village Wangi, one active woman has organized four ROSCAs. One of these has recently abandoned the lottery system altogether in favour of a pure savings club. Members' savings are deposited at a local bank, and no longer circulate within the club.

The lottery with discount counted only a few devotees and then only among traders and shopkeepers in the urban centres. The following example comes from Ramanandnagar and explains its mode of operation*. The ROSCA had 12 members, a cycle of 8 days, daily subscriptions of Rs 100, and a discount of Rs 300. Each fund amounted, therefore, to $12 \times 8 \times$ Rs 100 = Rs 9600, minus the Rs 300 discount, which was to be distributed among the members who had not yet received the fund. The fund was therefore Rs 9300.

At the first meeting the Rs 300 were divided among all 12 members and each received Rs 25, while the winner received Rs 9300 in addition. At the second meeting the remaining 11 members including the new winner of the fund, received Rs 27

---

* Data are supplied by Jeske Kortenhorst and Kees Zevenbergen.

each. At the third meeting 10 members received Rs 30 each. At the twelfth and final meeting, the last winner received the full discount and pocketed a sum of Rs 9600. He had already received a total of Rs 630 in the earlier meetings, representing about 7 per cent interest on his savings of the preceding 95 days. This comes to approximately 27 per cent on an annual basis. Late winners always earn a higher rate of interest on savings than earlier fund winners, and the last one in the row gets the highest rate of all.

The ROSCA or rotating-type society has existed in Sangli as long as people can remember, at first with grain and later with money contributions. They are the older form of informal financial self-help organizations. They seem to disappear gradually, however, and make way for the RESCA, which is now the predominant savings and credit model of the informal sector in the district. This is probably due to the impact of the sugar and dairy industries on the economy.

These industries changed the economic pattern of the district, as well as the occupational pattern of ROSCA participants. This had a profound influence on the demand of financial services. 'Whereas the main purpose of the agriculturist subscribers in joining the *chitties* was accumulation of savings, the primary aim of the traders and merchants was borrowing' (Nayar, 1986: 34). It was then that the chitty's main shortcoming became clear, and that was the very mechanism of the lottery so dearly loved by many players.

It is true that a ROSCA allows members (that is, all except the last one) access to a lump sum of money earlier than when they would have saved individually. But because of the lottery mechanism, a player cannot be sure of the moment he will receive the fund and he will not get access to his own savings at the time he may need them most. This is particularly frustrating when one has the chance to invest in a profitable opportunity. Also, because only one player can be accommodated at a time, the ROSCA is less suitable when many members need the fund simultaneously for a specific occasion, such as a religious Hindu festival. Of course one could, as Nayar observes (1986: 34), manage his affairs in a roundabout way by borrowing the necessary money from a local moneylender and repaying it when one wins his ROSCA fund. But one could also try out a variation on the ROSCA technique; hence the RESCA. The RESCA is a self help organization that accommodates borrowers as well as savers and offers a satisfactory alternative to the moneylender. This is most gratifying in times of economic growth and increasing consumption and investment opportunities, when the demand for credit is sure to rise fast. These conditions explain why the RESCA has become more popular than the ROSCA in Sangli.

In Sangli the people use the term bishi to denote both the rotating and non-rotating models. To prevent confusion, the name bishi will be used henceforth only for the RESCA-type of society. This type of society will be discussed in the following paragraphs.

7.2 BISHI EMERGENCE AND RATIONALE

Like the UCS, the bishi in Sangli is a rather recent phenomenon. The oldest club was in the agro-industrial town Ramanandnagar and dated from 1965; in 1988 researchers discovered a bishi that had begun in Ramapur in 1960, when the Verala river had not yet run dry. All others were started after 1970. Bishi proved more numerous than UCS; in the ten communities surveyed in 1984–86[12]), 50 to 75 were counted as compared with 13 UCS. Their popularity is still growing among all strata of society. Two years later, a further growth of bishis was recorded in a number of the same villages. Their greatest concentration was in the commercialized irrigated areas, where villages often had five to ten bishis, in contrast to villages in the dry zones that had no bishis or only a few, and where contributions were smaller. This suggests a relationship with the revival and diversification of the economy, caused initially by the sugar boom and thereafter sustained by the growth of the dairy industry.

Bishis are popular because of their informal way of operating, the high return they offer on savings and the relatively easy access they provide to credit.

A bishi is primarily a savings club in which participants voluntarily join a scheme of contractual saving. From experience, people have come to realize that to save individually requires a measure of discipline that is difficult to maintain. A bishi therefore substitutes group saving for individual saving and puts pressure on lazy contributors with heavy fines that can increase to 15 per cent of a missed payment a time. Regular savers, on the other hand, can be rewarded with returns of 50 per cent and more on capital when loans are made from the bishi fund at a high interest rate. The importance of the savings dimension can be inferred from the starting dates or birth dates of bishis. These are also the dates on which the cycle of a yearly bishi ends and the accumulated fund is distributed. An example is the situation in Vasagade.

Vasagade is a rural community in Tasgaon taluka of roughly 6000 people and 800 households. Its agriculture is characterized as semi-irrigated; that is, half of its cultivated area has access to water. In June 1984 Gerner recorded the existence of eight bishis, with a total of 678 members. The oldest club dated from 1974, the youngest from 1983. Six of the eight bishis had been started in October/November and people said this was because of Diwali, the big Hindu festival in November. This is an expensive affair for which people customarily save during the year. It is for this festival that members had started their bishi and put aside a sum of money at each monthly meeting.

People also join bishis to be able to borrow small sums of money for a short term in relative convenience, that is timely, nearby and without much fuss. Policies and procedures of formal financial institutions render it almost impossible to get credit for solving sudden liquidity crises; therefore, even persons, with accounts with more than one bank also participate in bishis— even bank employees themselves.

Credit is especially important to business people. For them the annual cycle of a bishi, clearing all available funds each year and starting anew with an empty treasury, is an inconvenience. They try to overcome this disadvantage by joining several bishis. But when most clubs in a locality have the same birthdate as in Vasagade, there still is a money-less period with empty coffers when they have no convenient access to credit. Thus they may start a new club with another birthdate. This is probably why Hospes, visiting Vasagade in 1985, reports the existence of ten bishis as against the eight reported by Gerner in 1984 (Gerner: 68; Hospes: 82). But a more logical and frequently used solution is either to operate the bishi with a cycle of two or more years or to distribute only part of the accumulated fund of a one year bishi and retain the other part for lending purposes. Both types of bishi were found in the survey and were usually established by people who prefer the lending dimension of the bishi to the saving dimension: business people such as merchants, traders, shopkeepers, artisans, rich farmers and owners of a small industrial enterprise. The number of these enterprises is growing fast and with an astonishing variety: workshops to cast and polish (spare) parts of a pumpset; to make chalks for blackboards; to mold plastic bottles, caps and T-pieces for electrical conduits. In Ramanandnagar, the so-called king of bishis is a merchant; he is chairman of five bishis, one of which has a five-year cycle.

7.3. COMPARING BISHI AND UCS

Bishis differ from UCS on several points. First, bishis are smaller and do not require an initial membership of 200. Organizers put the ideal size between 12 and 50 members to keep the organization manageable and social control effective. Many clubs, however have grown very fast and count between 60 and 80 members, while a few have more than one hundred. It is quite possible that such groups will become unwieldly and will split up eventually.

As a small club, a bishi is homogeneous. Members usually belong to the same caste and income group, the rich and the poor participating in separate clubs. Over 90 per cent of membership is male. In some bishis, age or profession is a selection criterion. Narsinhpur, a fully irrigated village of 4000 population in Walwa taluka, has no UCS but boasts several bishis. The oldest one was started in 1978. In 1985 it had 45 members; 30 are college graduates of different disciplines who preferred to go back to farming rather than trying their luck in a new profession. The other members are employees of the nearby sugar factory. All members are between 20 and 40 years of age.

In Takari, Walwa taluka, 25 milk collectors started a bishi in 1984 with weekly contributions ranging from Rs 25 to Rs 100 per member. Each week a loan of Rs 1,200 is given to one member in turn, which he usually passes on to a farmer owning cattle, in order to gain his favour. The latter may use the loan to buy a new buffalo. The milk collector takes repayment in fresh milk that he sells to the local dairy factory for a profit of Rs 0.50 per liter. The members use the bishi mainly to please their farmers-clientele and increase their milk sales to the factory.

Another such bishi was started by 12 milk collectors in Ramanandnagar in 1984 for the same purpose (Van den Boogaard: 86).

Second, contributions in a bishi are periodic, while in the UCS many participants do not save beyond the initial payment of share capital. Bishis have monthly or weekly meetings with individual contributions ranging from one to several hundred rupees per meeting.

A regular income is therefore required for membership. This is why bishis flourish in an urban environment that offers jobs in trade and commerce, small scale industry, and government. Agriculture does not provide a regular income, but many cane growers in Sangli have additional income from dairy cattle or a side job in trade and industry.

Third, while a bishi is an organization of short duration, a UCS, once registered, becomes a permanent affair until liquidated by the Co-operative Registrar. Observers of rural institutions have seen the short duration as the great weakness of this and similar self-help organizations, but in practice it also has advantages. ROSCA and bishi do not 'die', but are reconstituted time and again. This has several benefits. After each cycle, dissatisfied members can quietly take their leave; the organization can rid itself of incapable managers and troublesome members and even dissolve the bishi for good, without causing embarrassment and loss of face. The author discovered earlier bishis that had been dissolved peacefully because of economic decline or a loan committee that practised favouritism.

Liquidation of a 'permanent' organization like the UCS or expulsion of its members is more cumbersome and has to follow procedures given in the by-laws. This may involve convening a general meeting, having a vote and needing a majority in favour of expulsion. Such public discredit involves loss of face; this may be one reason why in some situations no action is taken and unqualified but influential community leaders retain a seat on the board of directors to the detriment of the organization.

Moreover, a bishi can also use its rebirth for a change of policies, rules and procedures. Many bishis, in fact, have done so. Its *ad hoc* character, far from being a hindrance, instead makes a bishi adaptive to a changing climate and accounts for its very flexibility.

The one great disadvantage of the short life cycle, particularly to regular borrowers, is the distribution of all accumulated money at the end of the agreed cycle. To acquire a new fund, the bishi has to start all over again and cannot accommodate loan demands for a length of time. This is why many bishis have started to change the annual cycle into a cycle of five years. The UCS, provided it is well managed, can satisfy borrowers more readily because its fund is not totally depleted regularly; moreover, the UCS can borrow outside money from the District Central Co-operative Bank. The latter facility is the main reason for a bishi to aspire to UCS status.

In Narsinhpur, the bishi of former college graduates has not distributed its fund for years but has deposited Rs 60,000 with the local bank instead. When a member wants a loan he gets a cheque drawn on the bank. The present loan limit in Rs 1500 per member, so the fund is sufficient to accommodate all loan demands. Yet, saving is still continued, while the members consider the possible conversion of the bishi into an UCS. The greater lending potential of the UCS appeals to the more progressive members with future investment plans. The chairman of the bishi is a progressive farmer. He has, besides his cane and rice fields and local cows, a number of grade cattle and crossbreds, poultry and a biogas tank for the supply of home cooking gas. The nearby sugar factory acts as a catalytic agent for development activities; the local PACS offers many extra services such as a collection centre for the cooling and subsequent transport of milk to faraway Bombay (600 km). Yet, the chairman confesses to some hesitation, wavering between the advantages of a long-term bishi and those of an UCS. He is well aware of the failures of UCS elsewhere. The requirement of an initial membership of 200 renders, in his opinion, a UCS unwieldy. 'The ideal size of a bishi is 50 members'. The postulate of a limited size for proper management is also the reason why he disfavours an amalgamation of Narsinhpur's present four bishis: 'The other three groups are not run properly'. Both a large bishi and USC will facilitate factionalism and the emergence of rivalling sub-groups.

An original solution to this dilemma was presented by a bishi in Sangli town that has converted into a UCS while its core members continued their former affairs within the new structure.

The great difference between bishi and UCS is in the degree of autonomy. A UCS is registered under the Co-operative Societies Act. Its by-laws have to abide by the official guidelines and its registration is subject to conditions set by the Commissioner for Co-operation. As a financial institution it is also subject to the monetary regulations concerning interest rates. A bishi, on the other hand, is free to set its own policies and procedures according to circumstances and needs. This makes the bishi a much more flexible organization with characteristics and peculiarities to suit every and any socio-economic environment. Its high interest rates attract many savers without detering potential borrowers.

7.4 BISHI OPERATIONS

The basic modus operandi of a bishi is as follows. The first part of each weekly or monthly meeting is used for the collection of savings (and loan repayments when in order). Each member has agreed to a regular subscription until the end of the cycle; the size of these subscriptions may vary among members. There are minimum and ceiling contributions. Apparently this is to prevent a single participant from gaining too much influence, as occurs in the UCS. Late contributors are fined immediately.

Next, members submit their loan requests; each bishi has a loan committee of five to ten persons to decide on borrowing priorities. Not all bishi want borrowing to start at the very first meeting, some have a time lapse of several meetings to allow the fund to grow sufficiently and test members' ability to contribute before they obtain credit. Competing loan demands can cause problems, especially when the bishi is still young and its capital small. For this reason, board and committee members in some societies have agreed not to borrow themselves during the bishi's infancy; in a more-year bishi, this abstinence might even last one year. Other, and influential, committee members are less altruistic and may manipulate loan decisions in their own or their protégées' favour. Alternatively, one of the safety mechanisms of many bishis is that a (small) part of the fund is kept in reserve at a commercial bank to meet individual crisis situations.

Loans are granted for all kind of purposes: contrary to formal financial convention, emergency cases have priority. Sometimes a rich member is allowed to borrow a very large sum. The survey counted a number of loans of several thousand rupees for trading capital and the purchase of cattle. It demonstrates the function of bishi as a financial intermediary between many savers and a few borrowers. These savers are attracted by the high reward that a bishi offers them. Each bishi must have a proper mix of savers and borrowers, a club of only savers or borrowers would have problems to survive and have less appeal.

A bishi lend for very short periods of one to three months; those with a life cycle of more than one year allow longer periods. Interests rates are high and increase over time to stimulate early repayment so that other borrowers can be accommodated. Typical rates are 3 per cent for the first month and 5 per cent per month thereafter. Renewal of a loan is possible but not encouraged; higher interest is charged on renewals and must be paid in advance. Few members judge these leading rates too high. After all, when the fund is redistributed at cycle's end, members' savings are returned and the accumulated interest and fines are distributed. Borrowers therefore not only receive back part of their own interest payments, but also share in the (high) profits made on loans to other members.

All these safety mechanisms—bishi's small circle of intimates, the limits on contributions and advances, keeping money in reserve for emergency loans, the short-term lending against high interest—are intended to minimize chances of misconduct. Bad debt losses are reportedly almost non-existent. Late repayment of loans is not uncommon and punished severely; overdues may draw a fine of 10 per cent of the amount overdue. Savings, combined with interest payments and fines, make the bishi funds well considerably in a short time. All participants, savers in particular, reap the benefit when this fund is distributed at the end of each cycle; returns on capital of 30 to 60 per cent are recorded.

To give an impression of the magnitude of bishi funds, Gerner estimated the accumulated capital of the 678 members of the eight one-year bishi in Vasagade at Rs 342,000 at the cycle's end in 1984 (Gerner; 69). This was more than ten times the accumulated capital of the local, malfunctioning UCS. Vasagade

THE BISHI, A SELF-HELP ORGANIZATION 63

also has two PACS, one with membership of 500 and the other with 350. These societies do not collect savings. The first one disbursed loans of Rs 413,000, the other disbursed Rs 346,000 in the 1983-4 season.

Another and more telling illustration comes from Kundal and is reported by Hospes (80-82). Kundal is a small town near Ramanandnagar, the site of the Kirloskar agricultural machinery factory. Its population of 13,000 consists of factory workers, farmers and an unusually large number of merchants and shopkeepers. During the colonial days the town had a reputation for political agitation and anti-English sentiment. Its people are still very active in politics and in the promotion of co-operatives—which, in India, seem to go hand in hand. In Kundal there are two (co-operative) banks, a PACS and two dormant UCS, strangled by political rivalry (cf. Chapter 6, page 49).

Kundal boasts at least 10 bishis. One of them was founded in 1970 and grew steadily to 120 members in 1985. Membership in this bishi reflects Kundal's population, a mixture of farmers, factory workers and small businessmen, sometimes combining all three professions. Weekly contributions are between Rs 20 and Rs 600 per member and total Rs 9000 per week. This one bishi, therefore, collects a sum of Rs 450,000 per year! This is more than the accumulated capital (savings plus interest on loans) of Vasagade's eight bishis combined. Loans are for two months only and the monthly interest of 2 per cent had to be paid in advance. The accumulated interest in the six months between August 1984 and February 1985 was Rs 18,444; fines for late contributions and overdue loans totalled Rs 2000 over the same period. The bishi had an account with the local co-operative bank where it deposited its funds that were not loaned out.

Because bishis have no office, no paid staff and the very minimum of paperwork, they operate even more cheaply than the other two grassroots organizations, the PACS and the UCS.

Variations on this basic pattern of meeting cycles, saving and borrowing, reward and punishment, are endless. In one bishi, for example, 28 members saved Rs 10 each for 15 months, during which time no loans were issued. Thereafter, the process was reversed and contributions stopped while loans were made to members and non-members alike. While the original fund was retained, only the accumulated loan interest was distributed each year (Hospes: 84).

Nagarale, a small village in irrigated Tasgaon taluka, started a commodity bishi in 1983, in which members received, instead of money, a set of steel pans at the cycle's end.[13] Its success was such that it has grown rapidly into an informal banking institution serving nine villages. In each village a trained and salaried employee collects savings on a daily (or less frequent) basis from 2500 members, and distributes and recovers loans daily. Surprisingly, while the majority of membership of bishi and UCS usually is male, here 65 per cent are women who save a few rupees daily from the sale of milk and eggs. Interest on savings is fixed at 10 per cent per annum. Loan demands are discussed in daily meetings; a shortage of funds in one village can be met from a surplus in another. Like other bishis in the survey, this one also uses the nearby District Central Co-operative Bank for the safekeeping of funds not loaned out to members. The maximum loan limit is Rs 5000, the maximum loan period six months. A novelty is the setting aside of an emergency fund to lend to members as well non-members in case of mishap. Another novelty is refusal of membership to politically active persons, undoubtedly inspired by bad experience with politically infused PACS and UCS.

There are two notable developments in bishi culture. One is a growing appetite for some UCS formalities. Examples are the issuing of passbooks to members for recording savings and loans; the use of printed forms for loan demands and accountancy (bishi bookkeeping records are available in bookshops); the printing of bishi regulations, which otherwise are customarily written down in a notebook, signed by members and kept with the secretary; and the requirement that borrowers produce two co-signers to guarantee a loan. There are even bishis that aspire to UCS status in the future. Secondly, there is a budding desire for a linkage with a formal banking institution for the safekeeping of bishi funds. Reflecting this desire, three villages situations exist in which the chairman and treasurer keep a joint bishi account at the nearest bank. Informants see the very informality of bishi as the main cause for this development. 'Compared to the UCS, a bishi is a less safe place for your money because defaulters cannot be sued'. This statement is all the more remarkable because default is reportedly almost absent in bishis but very much present in UCS. Also, UCS are noted for their reluctance to sue defaulters, and court cases take years to be resolved.

7.5 SUMMARY OF BISHI CHARACTERISTICS

Summing up, a bishi is an informal self-help institution for saving and credit. In Sangli, its emergence seemingly coincides with the sugar boom; similar to its formal counterpart, the UCS, bishis are most numerous in the irrigated zones of the district. They outnumber UCS and PACS combined, demonstrating their popularity. Apparently they offer savings and credit facilities not found elsewhere. The demand for such facilities has increased with the economic development of the district.

A bishi has a number of characteristics that cause it to function smoothly and efficiently and explain its popularity.
1. Membership is small and homogeneous, which makes the bishi manageable and controllable and avoids factionalism. This enhances social and organizational viability.
2. The bishi is a fully autonomous organization, unencumbered by the rule-making state bureaucracy. The freedom to make one's own rules familiarizes membership with the operational procedures and reinforces social and organizational viability.
3. Bishis are semi-permanent organizations that are reconstituted periodically. Continuous re-cycling allows correction of deficiencies and adaptation of operations to environmental changes. This makes the bishi a most flexible institution.
4. Smallness and autonomy produce a veritable kaleidoscope of bishis. Each village may offer its own choice of clubs to suit every taste and purse. Multiple membership is possible and increases access to services.
5. As a self-help institution, the initial emphasis is on savings, discipline and low-cost efficiency; all of which enhance economic viability. Loans are small and very short term and are for production as well as consumption and social security purposes; emergency cases earn priority. Bishis thus complement formal financial institutions, that are not interested in this type of lending.

6. Transaction costs for members in a bishi are low because of proximity and the absence of formalities. At the same time, this keeps operation costs of bishis low, too. Proximity facilitates credit rating of borrowers and this decreases the chances of default. Loans, however, are priced high, but borrowers are partially compensated by the high interest on their savings deposits, which share in the profits of making loans to other members. People in Sangli apparently are less concerned than their government with the necessity of low interest rates.
7. Several bishis show a trend towards copying UCS formalities that promote security: passbooks, printed rules, co-signers for loans. Lasting experience with PACS and similar co-operative organizations has made it possible to select those procedures and regulations that are worthy of imitation.

Bishis also have their shortcomings. Saving in a bishi requires a regular source of income that is not readily available in an agricultural economy, Savings, once deposited in the bishi, are not liquid until the end of each cycle, unless the saver applies for a loan. Because of its short life cycle, the credit potential of a bishi is very limited and particularly so at the beginning; it is only near the end of the cycle that competing loan demands are met. Further, loan demands can only be met on meeting days when members convene; bishis do not provide 'instant' credit. Finally, bishis offer no secure haven for one's savings; bishi fraud or default can not be contested in the official courts. Still, the many advantages outnumber the disadvantages of this popular self-help organization.

7.6 RECENT DEVELOPMENTS IN BISHI CULTURE.*

A follow-up survey in four communities in 1988 revealed a further growth of bishis and a number of improvements in bishi financial technology that were apparently intended to remedy some of its more evident shortcomings. This survey thus confirmed the earlier impression of a versatile and adaptive self-help institution in the informal financial market of Sangli.

The follow-up survey discovered bishi that, in addition to

---

* Research data come from Jeske Kortenhorst and Kees Zevenbergen.

periodic subscriptions, collected large-sized one-time lump sum payments in order to boost a society's working capital. Such contributions were called 'shares', a term apparently borrowed from the co-operative culture of the formal finance sector. When 'shares' are solicited at the start of a new club, they largely solve the capital shortage problem of the society in its infancy. Without 'shares' contributions, clubs might have to postpone borrowing by members for fear of not being able to meet competing loan demands without quarrels. 'Share' contributions will be particularly welcome in a rural economy, where erratic income flows make substantial weekly or monthly subscriptions awkward, but allow sporadic and more sizeable contributions.

Another development is also indicative of the growing importance of the lending dimension over the savings dimension of the bishi. Bishis with a life cycle of one year have a limited credit potential, which inconveniences the more regular borrower. This has evoked a trend towards a longer duration, already noted in earlier surveys and confirmed in 1988, particularly in the commercialized communities in the sugar belt. The five-year bishi has become the prevailing type of savings and loan society in busy **Ramanandnagar** and in fully irrigated Narsinhpur and Ankalkhop. This means not only that the bishi fund, rather than being depleted every year, continues to grow for five years to meet more demands; but also that the bishi's ever-swelling capital will allow larger and longer loans. While the short-cycled bishi lends for periods of one to three months, loans from the five-year societies are commonly for a period of six to twelve months or longer. Even cases of loans of the duration of the bishi have been noted. A sizeable long-term loan at the start of a new business venture can make the difference between success and failure.

Of course, long-term bishi loans also carry a higher rate of interest (cf. section 7.4). This may inspire members, acting in concert with the board, to try the ruse of 'rolling over', in which the same loan is 'paid back' and immediately reborrowed each month. This is wholly a paper affair to avoid the higher interest rate; it smacks of the favouritism practised in the formal finance sector and is resented by most bishi participants. Eventually it may be the reason that loan requests of other members are postponed or not honoured, and may lead to the bishi's demise.

There are always exceptions to the rule and times when a large part of the bishi capitals lies idle for lack of loan demands. This may happen in the period when the payments by the sugar factory for cane deliveries reach their peak. In such instances, members who join a bishi primarily because of its high return on savings will have few objections to other members who deplete the fund and are thus largely responsible for the bishi's handsome dividends.

Savings invested in a sewing machine, an ox-cart, sheep, a cow or a buffalo will produce new income from tailoring, ox-cart rent, wool, milk and cow-dung and may whet the appetite for further advancement. Today's savers may then eventually become tomorrow's investors seeking new venture capital. As long as Sangli's economy shows signs of continuing growth and diversification, the lending dimension of bishi will be increasingly appreciated.

Another recent development may stem from this growing importance of the bishi's lending dimension. Many societies nowadays no longer hold monthly meetings to which members bring their contributions and participate in the social life of the bishi. Instead, members simply hand over the contribution to the secretary, at his house, during the first week of each month. At the end of the month, the loan committee convenes to decide on the loan demands that have meanwhile been submitted.

All bishis in Ankalkhop and Narsinhpur appeared to function without regular meetings. It was argued that multiple membership of bishis in some instances and nightshifts by workers in the sugar factory in others, were apt to make attendance at every meeting too troublesome and time-consuming. For whatever reason, it is clear that the bishi's social function has been eroded, at the same time that its economic function as a financial institution has gained importance. The same development has been noted in ROSCA-culture under similar conditions (Bouman, 1979, 1984). It harbours the danger of a break-down of the bishi's social control mechanism and a subsequent increase in abuse and default, which will make new rules and control procedures necessary.

Declining social control and more complicated administration of a five-year bishi imply increased responsibility and work for bishi functionaries. Handling a society's affairs may become too burdensome and thus force a secretary to resign. This has been the case with the successful bishi of former college graduates in Narsinhpur. The chairman, who had been at the helm since the start in 1978, resigned in 1987 when the bishi had completed its five-year cycle and had amassed the impressive capital of Rs 116,000. The bishi promptly folded, and a new one has not been started. Conversion into a UCS may still be debated (cf. section 7.3).

The emergence of the lending dimension as the prime attraction and major activity of a bishi, with its potential consequences for management and social control, may have hastened the shift towards copying formal-sector formalities that was noted earlier (cf. section 7.5). These formalities, such as passbooks, printed by-laws and standard bookkeeping forms, co-signers or collateral (jewellery, gold) for loans, promote greater security. Finally, most bishis nowadays appear to have links with commercial banks, where funds are deposited in a joint account of two dignitaries, and loans are granted by writing out checks. Besides offering greater safety and convenience, bank deposits have the additional advantage of being insured by the state. Usually there is some money left in the account to accommodate emergency demands for a loan; hence, membership in a bishi offers the possibility of access to instant capital.

CHAPTER 8

Of moneylenders and pawnbrokers

In his appraisal of the position of the moneylender, Franda (p. 39) quotes an old Indian saying that a village fit to live in must have a moneylender, a medical practitioner, a man of knowledge and a stream that does not dry up in the summer. The significance of the moneylender for the continued existence, if not the welfare, of the rural populace could not be expressed more clearly.

In the literature on rural development the term moneylender has many connotations. Moneylenders come in all shapes and sizes, from the evil local Scrooge to the shopkeeper, landlord, miller, secretary of the co-operative, village headman, fertilizer dealer or the produce buyer. Strictly speaking, few are professional moneylenders but practise lending either to promote their business as produce buyer, sales agent or processor of goods; or to protect their reputation as administrator and office bearer. Some lend money only occasionally; some advance credit in kind. Here, the term moneylender will be used literally: to lend money and not grain, fertilizer or any other product, and not occasionally, but as a full-time professional. The pure 'moneylender' is gradually disappearing from the rural scene, but as a pawnbroker he is still very much in business.

One may lend money on personal security, that is without collateral, or on the tangible security of a pledged item. The former is risky and requires close knowledge and supervision of borrowers. Such credit is advanced only to personal acquaintances, and the lender can serve but a few clients unless he employs a network of intelligence agents. Lending on pledged sureties reduces risk and the necessity of close contact and surveillance, and therefore allows a much larger circle of borrowers. The items that are pledged can be anything; a watch, bicycle, radio, cloth, utensils. Much more common is the pledging of valuables like jewellery and gold or silver ornaments. Moneylending of this kind is known as pawnbroking and is an attractive way of lending money because of the low risk and transaction costs. A loan transaction takes only a few minutes, the time necessary to appraise the value of the pawn. There is no need for the lengthy application forms that a bank normally uses when assessing credit-worthiness of a prospective client. Assembling information on repayment capacity and trustworthiness, as well as drawing up and registration of mortgage deeds become superfluous. Loan administration is simple and straightforward and procedures can be kept to a minimum. Pawnbroking is also attractive to borrowers who need not incur additional obligations such as buying from or selling produce to the lender. The borrower simply sells an item to the pawnbroker on the condition that he may buy it back within a specified time for the same amounts plus interest. If he does not buy back, the broker may sell or keep the item for himself.

While the relative importance of informal lenders in developing countries has declined in recent years with the expanding role of formal financial intermediaries, the business of the pawnbroker flourishes, possibly because of the steep rise in the gold price since 1971. This chapter examines pawnshops and pawnbrokers dealing in gold and other valuables.

8.1 GOLD AS EVERYBODY'S PIGGYBANK

The storing of valuables and, in particular, the hoarding of gold, have always had an immense appeal to savers. Unlike the coins and banknotes of an individual country, precious metals and stones have universal worth. They have a high degree of liquidity because they can be sold or pledged for cash almost anywhere. Drake, refering to Malaysia where buyers of gold articles may sell these back to the shop at around 90 per cent of their retail price, points out that 'the existence of pawnshops and of sale-repurchase practices among jewellers is of crucial importance in conferring this liquidity' (Drake: 125–6). Moll (1989) found the same type of arrangement in Indonesia in 1986. Further, jewellery offers emotional satisfaction and bestows status on the wearer; valuables also function as a hedge against inflation, while bank balances are eroded by it. Gold and stones have a high value in relation to their bulk and are easy to hide, transport and negotiate. This is important in times of natural disaster, war or social unrest. In Sri Lanka, even poor Tamil plantation workers have historically converted their savings into valuables for this reason (Bouman and Houtman: 73).

It is important to recognize the role of gold in fostering the savings habit. Economists may find the hoarding of gold a wastefully unproductive use of savings. They much prefer people to save in the form of financial assets such as deposits with banks, for a better allocation of resources. However, as Drake points out, saving in gold may not be deleterious to economic growth. What matters is the ultimate use of funds obtained by the seller of gold or by the borrower who pledges it with the pawnbroker (Drake: 124). They may use these funds for consumption as well as for productive investment and capital formation, as reported by Harriss (1981: 165, 195), Sivakumar (1978: 847) and van Nieuwkoop (1986: 63, 74).

In Asia, saving in gold has historically been popular among all strata of society. In some parts of Asia private gold hoards are estimated to equal 10 per cent of national income (Drake: 125, quoting Gamba). Bouman and Houtman report widespread demand for gold in Sri Lanka. In a 1986 study of the economic behaviour of vegetable growers in West Java, Indonesia, Moll found that the acquisition of gold ranked second in investment priorities, immediately after investment in the farm business. These farmers also reported that pawning with the local pawnshops had become a matter of routine in their daily life (Moll, personal communication). But perhaps the clearest case of gold fever is India, where gold smuggling across the Gulf is an ancient art.

In a survey of two villages in Tamil Nadu, in 1976, Sivakumar found that the value of gold stores of all categories of peasants for exceeded the value of their cash and bank balances. He observes that 'all classes possess gold and silver, even the landless peasants ..... The common practice of dowry consisting of gold, silver and brass-ware must be seen in this light' (Sivakumar: 847). According to Nayar (personal communication), in Kerala gold occupies a place second only to land and buildings in the asset structure of households. The late Indira Gandhi indicated the importance of gold for the ordinary people of India: 'India is not a poor country if you count the amount of gold. Each woman owns at least one tola (11.66 grammes) of gold ornaments. There are more than thirty crores (300 million) women in India ......' (quoted from Bastiaansen: 1).

Gold's continued popularity in the country becomes understandable in the light of the price of gold in India, which apparently behaves independently of the world market. From Rs 87 per 10 grammes in 1946 it has climbed continuously to Rs 2200 in April 1986. Its value increased almost 20 per cent between June 1980 and June 1981; this contrasts with a sharp fall of 36.4 per cent (from $ 662.50 to $ 421.50 per ounce) in the London market during that period. According to a recent report, the great demand in India has pushed local gold prices about 40 per cent higher than on international markets and stimulated the smuggling of 100 tons of gold into the country in 1987 (Newsweek: 21).

Table 6. Gold prices in India (Rs per 10 grams)

Year	Average price	Year	Average price
1946–47	86.66	1975–76	519.10
1950–51	97.28	1976–77	549.50
1955–56	82.18	1977–78	637.33
1960–61	114.91	1978–79	994.64
1965–66	133.34	1979–80	1345.00
1970–71	184.96	1980–81	1610.00
1971–72	200.16	1981–82	1655.00
1972–73	242.14	1983 (Febr.)	1740.00
1973–74	369.23	1984 (Febr.)	1895.00
1974–75	519.10	(July)	1985.00
		1986 (April)	2200.00

SOURCE: C.P.S. Nayar (Personal Communication)

When the USA abolished the official gold price of US $ 35 an ounce in August 1971, gold started its spectacular climb in the free market, rising to $ 850 an ounce in January 1980. Since then it has come down almost as spectacularly and has fluctuated between $ 300 and $ 500 an ounce. Such wild movements not only attracted speculators, but also prompted a renewed worldwide interest in gold as a store of value for the regular saver. Saving in gold has become a common phenomenon and wherever this is the case, the pawnshop business flourishes.

The increase in public and private, licensed and unlicensed pawnbroking has been global in character. Pawnshop stories, anecdotes and dramas appear regularly in the popular press of both industrialized and developing countries.[14]

8.2 GOLD AND PAWNBROKING IN ASIA

Most reports of pawnbroking activities come, again, from Asia. In Indonesia, where pawnbroking is a legal monopoly of the government, the growth of pawning has been impressive, from a volume of 31 billion rupiah in 1975 to 156 billion rupiah in 1981, which represents an average annual increase of 32 per cent. Redemption exceeds 95 per cent. The government pawnshops are reportedly more effective in giving the poor access to institutional credit than the many subsidized credit schemes of the State Banks (McLeod: 95). In addition to these government shops numerous illegal private houses exist, doing a brisk business in towns and in the countryside.

Wells studied the credit arrangements of irrigated paddy farmers in two separate surveys in Kedah State, Malaysia, in 1972 and 1980. He found that pawnshops provided 19 and 18 per cent respectively of total borrowings and came second after shopkeepers as a source of lending within the informal credit market[15] (Wells, 1979: 92; 1980: 20). Farmers pawned gold and jewellery during the crop season and reclaimed them after the harvest. Average nominal interest rates in both years were 14 per cent for six months or 2 to 3 per cent per month, depending on the amount borrowed, with the smaller amounts carrying the higher interest rates. Contrary to Gamba, Wells found no evidence of monopolistic exploitation or excessive profiteering (Wells, 1980: 29).

Six years later, in the national Malaysian rural credit survey of 1986, pawnbroking emerged as the most important source of informal lending in Malaysia's rice producing areas, both rainfed and irrigated. Fifty four per cent of the paddy farmers who borrowed informally took their loans from pawnbrokers.[16]

Most of these advances (over 50 per cent) were used to finance cultivation expenses and education fees (24 per cent), and were repaid after the harvest (van Nieuwkoop: 63–4). The interest rate was still 2 to 3 per cent a month, with borrowing periods stretching to six months. According to van Nieuwkoop these loans competed directly with the heavily subsidized loans of the Bank Pertanian under the paddy production scheme. The Bank Pertanian is the State Bank officially charged with giving credit to promote agricultural development. The number of private pawnshop loans by far outnumbered the Bank's short term loans (van Nieukoop: 74).

Bouman and Houtman (1988) give some details of pawnbroking in Sri Lanka. Here, the rural offices of the People's Bank, the national bank in charge of financing the rural development programmes of the government, have engaged in pawnbroking since 1964. Pawning facilities of these banks gradually became so important that in 1981 80 per cent of all rural banks' lending was based on pledged gold and jewellery. Still, private houses carried on a brisk business[17] and boasted a larger patronage than the rural offices of the People's Bank. Turnover of the average private broker was 10 to 20 times that of the average rural bank. Yet, private houses did not serve customers indiscriminately; instead they exercised careful selection, for fear of accepting stolen goods. They also preferred borrowers to redeem pledged possessions, to avoid the need to sell the property at an auction.

Due to the wildly fluctuating gold market, loan periods in Sri Lanka were extremely short, varying between two weeks and two months. The majority of loans had a value between Rs 1000 and 2500, equivalent to US $ 50 and 125 respectively. Informal co-operation existed between private brokers, the licensed broker refinancing the unlicensed, who acted as his agent. Interest rates of private pawnshops were 4 to 7 per cent a month, greatly exceeding the bank rate of 30 per cent a year. In part this is because the shops are charged a tax of 2 per cent on turnover by the government which, of course, they pass on to the customer. Despite these higher charges, the public patronizes private brokers because of the superior service they offer. Calculation of pawnshop economics revealed no huge profits or excessive returns on capital; like Wells in Malaysia, the authors found no evidence of monopolistic behaviour of these private lenders in Sri Lanka.

8.3 THE INDIAN EXPERIENCE

The Indian literature is more informative than any other on the subject of moneylenders. Perhaps this is because the country historically can boast of 'dynasties' of mercantile lenders, organized along lines of an ethnic community, such as the Nattukottai Chettiars of Tamil Nadu, the Marwaris of Rajasthan and the Banias of Gujarat. These ethnic communities have for centuries acted as bankers to Maharajas as well as to humble peasants (Franda: 39) and have become the subject of much heated debate: should they be considered agents of change and development or of stagnation and underdevelopment? Even Darling's classic tale of the Punjab peasant in prosperity and debt (1947, reprint) breathes ambiguity: yes, they are rascals, but no, the country can't do without them. Of the more illustrious lenders, the Chettiars ventured overseas and spread out over the whole of Southeast Asia, while the Marwaris remained in India and settled 'in every conceivable nook and cranny' of the country (Franda: 40).

The moneylending profession proper seldom enjoys high repute anywhere. This is why outsiders, ethnic or religious usually take up the trade initially, to be followed by locals who may start as their brokers or agents and may be refinanced by them. Besides lending, they may also accept savings and deposits of clients, for which they may engage collectors (Timberg and Aiyar: 280, 283; Harriss, 1981: diagram 167). The moneylending dynasties created an impressive network of indigenous bankers in India with a high degree of organization and self-regulation through local and regional associations (Timberg and Aiyar: 285).[18]

Pawnbroking is almost always combined with goldsmithing, gold refining and jewellery; lending on pledged valuables has been an essential part of the ancient business of reputable professional lenders. Lately, however, the stage has been filled with other actors. Harriss notes a massive increase in the number of pawnbrokers throughout the North Arcot District of Tamil Nadu betwen 1965 and 1974 (when she finished her survey).

From 10 in 1965 to 72 in 1973, the number rose to 88 in 1974, most of the new entrants being South Indian castes in contrast to the Marwaris, the outsiders of former times. The author remarks that this is a trend common all over the District (Harriss, 1981: 166). Besides this intake of new pawnbrokers, the District also saw the emergence of minor government officials, teachers and clerks as a new class of non-professional lenders, a phenomenon also observed by Kurup (1002) in Kerala. Elsewhere, this new class may consist of successful farmers who have benefited from the Green Revolution. The new lenders operate only partly with their own funds; many of them also borrow from banks and professional moneylenders for onlending.

Harriss (1981: 166) explains this sudden swelling of the ranks of pawnbrokers and moneylenders from an increased demand for small agricultural loans, a demand that trebled in 1972–4. Others see it as the result of greater restrictions imposed on the profession by the various Moneylenders Acts and Debt Relief Acts. Referring to the Mysore Debt Relief Act of 1966, which scaled down farmers' debts with private moneylenders and provided for the ultimate transfer of the debt, if still unpaid, to a bank, a disbelieving Hill labels these regulations 'incomprehensible' (Hill: 219). Similar laws have been enacted in other States at other points of time, apparently to the same effect. Timberg and Aiyar (298) who did a comprehensive survey of the Indian informal credit market in 1978, note that Rastogi and Chettiar financiers, who once lent extensively to agriculturists, no longer do so, a development that they attribute to the "onerous regulations" of these acts. But it also seems quite conceivable that some of the moneylenders proper have made a tactical retreat and now content themselves with acting as refinancers to the new classes of pawnbrokers and moneylenders, to whom they leave the actual serving of the public.

Although according to Harriss the pawnbroker's clientele as a whole is not poor, pawnbroking is becoming important at the poor end of the peasant spectrum (Harriss, 1981: 166). The increasing demand for loans that she and others have noted, comes mostly from poorer farmers and is stimulated by the introduction of improved cultivation practices. Harriss mentions a demand for sums between Rs 50 and 100, small sums that traders (and, one might add, banks) will not lend (1981: 165).

Also Kurup (1002) and Sivakumar (847) concur that poor peasants pledge gold and silver, even brass vessels, to procure cash for cultivating expenses. Larger sums are sometimes borrowed for investments in irrigation, but also to cover ceremonial expenses and gambling debts (Harriss, 1981: 165; Sivakumar: 848). Although others have generally stressed the 'unproductive use' of pawnshop loans for household expenses (see e.g. Rozental for Thailand, quoted in Drake: 129), the very detailed Indian studies as well as those of Wells and van Nieuwkoop for Malaysia, confirm that valuables are pledged with lenders for on-farm investment as well. Of course, pawnbroking is not confined to rural areas; it is also prominent in cities and towns where the end use of borrowing has a completely different nature. The rough figures of Timberg and Aiyar (295) suggest this.

Harriss reports interest rates of 18–25 per cent on pawnshop loans in 1974, when relevant bank rates were 12 per cent. The highly seasonal demand for these loans (January to May) causes idle capital with the lender, and the interest rate must reflect this to some extent (Harriss, 1981: 166). She remarks that these interest rates are lower than in conventional characterizations of Indian agriculture (Harriss 1983: 237); Timberg and Aiyar (279) concur on this point.

Kurup is less explicit in quoting interest rates because he recognizes the possibility of hidden charges. The normal loan period granted by moneylenders is only three months, but renewal is possible after clearing interest arrears (1002). He mentions only one gold-based Rs 200 loan from a professional lender, who in 1975 charged 36 per cent (1005), which is apparently equal to 3 per cent a month. When Timberg and Aiyar studied the informal credit market in 1978, the bank lending rates were between 13 and 16 per cent per annum. Informal rates could go up to 36 per cent, with Chettiar pawnbrokers charging between 18 and 30 per cent (280). But the modal informal rate was 18–24 per cent 'or only 2 to 8 per cent higher than the bank rate' (287).

Sivakumar presents precise data from two villages in Tamil Nadu in 1976. He refers to the leading role in the district of the Marwari moneylenders, who arrived in the study area about a century ago. 'The ready availability of cash with the Marwaris and the ease with which loans can be obtained from them ensured the continued holding of gold and silver by the petty peasants' (847). Loans pledged with Marwaris are for six months, with monthly payments of interest. Interest rates vary with the value of the items: on gold objects, the rate is 2.5 to 3 per cent per month, on silver 4 per cent, on brass and copper 6 per cent. Rates may be up to 10 per cent on durable goods such as a watch, a torch or a radio (848). Risk is of course greatest with the latter items, and the loans granted against these goods will probably be very small, which explains the much higher interest rate.

In an intriguing study, Platteau and his co-researchers (1980; 1984) reveal the characteristics of moneylending in a poor fishing village in Kerala in 1978. Most borrowing is community based and credit comes from relatives, friends, shopkeepers or boat owners. Only a little more than 5 per cent of the loans, representing 7.5 per cent of total value, come from moneylenders and are obtained against the pledge of gold (Platteau et al., 1980: 1767). Usual interest rates on such loans are 18 to 36 per cent, but in some areas these rates reach 60 per cent. Why this occurs is not explained, except for the author's suggestion of scarcity of loanable funds or exploitive credit practices (Platteau, 1980: 61).

Authors reporting on pawnbroking activities in India are thus remarkably consistent about the interest rates that the private houses charge their clients: between 1.5 and 3 per cent per month, except for some extreme cases. This consistency is probably explained by active competition in the field. Pawnbrokers compete not only with each other, but also with the commercial banks. The latter lend money on pledged jewels and gold against lower rates of interest but restrict their services to more wealthy and middle class peasants.

Harriss' research convinced her of the absence of monopoly in the money market because the three necessary conditions for monopoly were lacking: 'small number of participants, protective barriers to entry or control over substitutes for money' (1981: 166). Timberg and Aiyar agree (301); their extensive study of the Indian informal credit markets also persuaded them that the actors in these markets play a positive role in the economy. They serve sectors and needs that banks refuse to serve. 'The rapid development of informal sources of credit to meet the demands cut off by bank selective credit controls again shows a response to demand' (Timberg and Aiyar: 298).

8.4 SUMMARY

To summarize this survey of pawnbrokers in Asia: in contrast to banks, they serve the poor as well as the rich. They grant loans for both production and consumption and do not charge the extravagant rates of interest of conventional characterizations of moneylenders. They compete with each other and with commercial banks and outperform the latter on many counts.

Several factors are generally mentioned that account for the superior performance of informal lenders. They have convenient opening hours, offer easy access and speedy service, have trained and experienced appraisers, avoid bureaucratic paperwork and embarassment, give larger loans in proportion to value pledged and do not place restrictions on the use of loans (Harriss, 1983: 235; Hill: 214; Oxfam: 18; McLeod: 328; Bouman and Houtman, 1988). Pawnbrokers may even accept deposits from the public which they invest in their business and re-lend to others (Timberg and Aiyar: 283). They are refinanced by both commercial banks and other moneylenders[19]. Thus, pawnshops are a promising point of integration between informal and institutional financing (Drake: 141).

CHAPTER 9

Pawnbroking in Sangli*

Pawnbroking is probably the oldest type of moneylending because of its comparative advantage over other forms of lending. This is especially true for India where the price of gold has increased continuously over the past 40 years (see Table 6). If the borrower does not return with the money to redeem his loan, the lender is practically sure to be compensated fully by the sale of the pledged gold object.

In India the pledging of gold and other valuables has been popular with people of all castes and classes, who have long become accustomed to save in gold, silver and brassware. Apparently lenders as well as borrowers agree on the convenience of pawnbroking. Yet, Harriss believes that pawnbrokers occupy a low social position. She finds that pawnbrokers in North Arcot, on the one hand, 'are a more accepted part of agricultural society than is the formal sector where people feel at a disadvantage'; on the other hand, 'they are held in low esteem by other lenders in the market' (Harriss: 166). Unfortunately the author provides no further explanation[20]. In Sangli, pawnbroking facilities are offered by agents in the formal and informal sectors of the financial market alike: commercial and cooperative banks, co-operative societies, licensed and unlicensed private moneylenders. Measured in terms of loan volume, the institutional sector is the larger one.[21]

9.1. PAWNBROKING BY FORMAL FINANCIAL INSTITUTIONS

Because gold advances usually involve small amounts, not all formal institutions like pawnbroking. This is manifest in the absence of a trained appraiser of valuables among bank staff.

---

* This chapter is based on research by Bastiaansen.

The commercial banks, practically all of them nationalized, are the least interested and often refer the small customer to co-operative banks, who cater to a clientele of lesser means. Some co-operative banks are known to specialize in pawnbroking and do have a trained appraiser to deal with gold advances. The Urban Co-operative Bank in Vita—a town that used to be a gold refining centre and still houses a great number of goldsmiths and jewel shops—did 27 per cent of its lending business in 1985 on the basis of gold pledging. At Ambika, the reputable UCS in Wangi, the proportion of gold pledging was as high as 40 per cent during 1981–4. Even some large private moneylenders are known to send small borrowers to such a specialized co-operative institution when it is located in the neighbourhood.

Repledging of gold offered by private pawnbrokers at banks, however, is another matter. Even though such repledging is unacceptable to the RBI, banks seldom turn away the private moneylender because repledging usually means big business and large loan turnover. Banks also may have a special relationship with a nearby jewel shop. A customer, a respected one in particular, who comes to pledge a gold object at a bank with no appraiser among its staff, is forwarded with a bank form to the nearby jeweller. The jeweller appraises the object and writes his estimate on the form. The customer then returns to the bank where he gets his loan. In case this jeweller (who also has his private pawnbroking business), comes to the bank to repledge the valuables that he has accepted from his own clients, the bank can hardly refuse to reciprocate a service. This arrangement works very well and to the satisfaction of both parties.

A commercial bank with a relatively large number of low income people among its constituency, may decide to open a special pawning window as the only way to reach the loan targets that the official Annual Action Plans have set for the bank. This is more likely to be the case in rural rather than in urban areas, because the prime target groups of these banks are the agriculturists.

A reluctant bank can also be persuaded by its clients to open such a window. This was the case in Burli in Tasgaon subdistrict (population 8,500) where local villagers requested the bank to provide pawnbroking facilities soon after it opened its doors to the public.

The RBI attitude toward pawnbroking by co-operative institutions is somewhat ambivalent. The RBI, which is the ultimate financer of the co-operative credit structure, originally disfavoured the provision of pawnbroking facilities by DCCB offices. This would lead to personalized retail loans rather than wholesale lending to PACS and 'would go against the structural discipline of the co-operative credit system'[22]. But in 1976 the RBI had to review its policy. Due to an unfair Debt Relief Act—considered a political gimmick by many—the credit supply of rural moneylenders largely dried up, without any other source taking its place. This credit had been extended mostly on the security of gold. Reluctantly, the RBI then permitted DCCB branches to grant pawnbroking loans to individuals 'for consumption purposes' provided such loans did not exceed Rs 1000 per borrower (successively increased in steps to Rs 5000 by 1985). In 1985 the gold advances of the DCCB head office in Sangli approximated 5 per cent of total loans. To summarize, the pledging of gold at formal institutions is more common in rural than in urban areas, more popular with co-operative banks than with nationalized commercial banks, and more logical for institutions that cater to the less affluent segments of society. Some UCS specialize in this type of lending, but PACS seldom provide gold advances. Both institutions need a special provision in their bylaws to offer pawnbroking facilities to members.

Lenders advance between 50 and 70 per cent of the retail gold value of the pawn. The difference appears to be a matter more of policy and cash position than of competition. The UCB in Sangli town lends against 65 per cent of value, Ambika UCS in Wangi discounts at 56 per cent, and both institutions are reputed to specialize in this sort of business.

Loan sums vary between the extremes of Rs 100 and 50,000. The smaller institutions such as the UCS usually apply a minimum loan amount of Rs 100, commercial banks one of Rs 300. Commercial banks commonly provide the larger advances of Rs 10,000 and over; these go to well-known trading firms and to moneylenders who have come for repledging. The average loan sum is estimated to fluctuate between Rs 500 and 3000 (Bastiaansen: 25).*

---

* The Bombay gold price (per 10 grams) was Rs 2200 in April 1986.

Usual interest rates vary between 16 and 17.5 per cent for both the co-operative and commercial banks. The latter, however, which as a nationalized institution have perforce become a public agent of development, can also apply the much lower concessional rate of 12.5 per cent to specific priority groups, indicated as such in the District's Development Plan. During his survey, Bastiaansen gained the impression that bank staff, who have no great love for this form of lending, did not adhere strictly to these concessional rates (Bastiaansen: 28). The Urban Credit Societies charged a slightly higher interest rate, 17 to 18 per cent, than their big brother, the Co-operative Bank.

As a general policy, a loan based on gold pledging has a maturity of six months, after which it is renewable. At some institutions the borrower has to pay one month's interest even when the loan is for a shorter period*, which is seldom the case. The pattern of loan terms is more or less as follows: 25–30 per cent of borrowers repay within three months; 70–75 per cent repay within six months; 90–95 per cent repay within one year. The remaining 5–10 per cent redeem only after two or more years.

The great majority of gold loans are, then, for a six-month term. Because of the steadily rising gold price in India, banks do not mind much when repayment is postponed, as long as the value of the pledged article covers principal plus interest. For the same reason, borrowers are eager to redeem. When a lending institution reminds a forgetful borrower that his loan has expired, he tends to respond quickly, and the parties try to come to some form of mutually satisfactory arrangement. Usually this implies payment of interest and renewal of the loan; if need be, the loan is renewed in the name of another individual to avoid violating the institution's policy not to extend gold loans beyond a certain period. Bastiaansen (28) was witness to such a loan extension involving the borrower's brother, who had come along to the bank. The whole deal was completed in one hour of cordial conspiracy.

The recent drought in the district that has already lasted for an uncomfortable couple of years, has forced many a loan extension. Repayment is expected after the rains have come. The pattern, to which lenders and borrowers have grown accustomed, is a

---

* This is generally standard pawnbroking practice everywhere.

familiar one in this semi-arid region of India. Public sales at which unredeemed articles are auctioned, are therefore a rarity. When necessary, a sale can be arranged by a reputable jeweller. As is common in the pawnbroking business, the proceeds of the sale go to the lender. After deduction of principal, interest and costs, any eventual surplus is paid to the pledgor.

One reason why borrowers, and in particular the ones at the lower income end of the spectrum, prefer the private moneylender despite his higher interest rates, is the lack of service offered by many a formal lending institution. To begin with, there are the inconvenient office hours of 10:30 a.m. to 3:30 p.m. Next there is the lack of experienced appraisers, so that a prospective borrower is sent to another address to get a valuation report. If he persists, he then becomes smothered with the love for institutional decorum in the Indian countryside. The number of forms that need processing is awesome. In one case ten forms were exchanged for a loan renewal. A borrower must open a savings account first, into which the loan is then deposited and from which it is subsequently withdrawn.

All these forms, naturally, have a price, however small. Taken together, they can make even a 'cheap' small loan rather expensive. The most common charges are the valuation fee of 1 to 2.5 per cent of estimated value, and payments for forms, insurance, stamp duty and membership or admission fee. Finally, some rupees have to be left in the (newly opened) savings account. If all charges are added, a borrower pays between Rs 30 and 75 for a loan of Rs 1400[23]. Transaction costs thus represent 2 to 5 per cent of the principal and must be added to the usual interest rate of 17 per cent to make a proper cost comparison between institutional and informal borrowing. Still, borrowing at a formal institution against the pledge of gold may be cheaper and a less harassing experience than borrowing money from a bank under other loan arrangements. This is especially true when rural borrowers get involved with concessional lending practices.[24] Of course clients who have frequent face-to-face contacts with bankers are less bothered by procedural extravagance. In institutions that specialize in gold advances and have their own appraisers who know their clients (and sometimes their jewels) well, business is conducted in a relaxed and friendly atmosphere.

Even if earrings, necklaces and other valuables commonly belong to women, the one who comes to pledge them with the broker is a male in most cases. According to the lenders, the majority of institutional pawnshop loans, and certainly those of lower income groups, meet consumption needs and social and ceremonial expenses. The pawning business peaks during March through May, the time considered auspicious for marrying. Gold ornaments are bought for the dowry and presented to the bride, with others to be worn by guests at the wedding ceremony. Indian marriages can be a very expensive affair and force many to pledge their jewels soon after having displayed them at the wedding.

Another pawnhouse pattern follows the crop cycle. As was observed by Harriss and Wells earlier, gold advances tend to coincide with the monsoon cropping season (June through September) and are used to meet farm expenses. Loans are also taken, but to a lesser extent, during the lean season that precedes the harvest, when food becomes scarce. Thereafter, from December onwards, loans get repaid.

Traders and moneylenders, too, are regular customers of the pawn credit section of the institutional lending houses. They are likely to invest the loan proceeds in their business. Private pawnbrokers come for repledging the valuables of their own clientele and to improve their liquidity position. Since this always involves large sums of money, the banks are happy to oblige; it is the smaller loan demands that they try to avoid.

9.2. PAWNBROKERS IN THE INFORMAL MARKET

There are two types of private pawnbrokers, licensed and unlicensed. Both are included here as part of the informal financial sector.

Strictly speaking, licensed pawnbrokers should not be counted as informal lenders. They are officially registered, subject to government supervision, and restricted by maximum interest rates. The licensed moneylenders* keep accounts in accordance with regulations; they are organized in the Sangli District Moneylenders Association.

---

* The terms pawnbroker and moneylender are used here interchangeably.

However, they are monitored and regulated much less than the commercial banks. The annual inspection by the Deputy Registrar of Co-operative Societies is superficial only; their interest rates may exceed the maximum permitted one; and in addition to pawn loans they give unsecured personal loans on a confidential basis. Following Timberg and Aiyar (Table 1, 280; 283) they are therefore classified as operators in the informal credit market.

As has been the case elsewhere in India, the early moneylenders in the district were outsiders (Franda: 39-40; Timberg and Aiyar: 285). They belonged to ethnic communities from Rajasthan and Gujarat; the Pathans even came from as far as Pakistan and Afghanistan[25]. Later, locals joined in and were refinanced by the largest among the immigrants, called *savkar*, which means 'rich man' in the Marathi vernacular. Gradually a network of professional and part-time moneylenders was established, serving kings as well as humble peasants and landless labourers. Money was lent on the pledge of gold, silver, copper and brass, while merchants also accepted grain and cloth, which were useful in their trade. The lenders also took mortgages on land and granted credit on personal security and third party guarantees.

Today, the moneylending business is about equally divided between locals and the offspring of the original immigrants.[26] The savkar, however, has more or less withdrawn from the business. Although a small moneylender may still refinance with a bigger colleague, the savkar's function of 'banking the bankers' is gradually being taken over by commercial banks—even when this runs counter to RBI instructions.

9.2.1. *The licensed pawnbroker*

As Harris has noted for North Arcot in Tamil Nadu, the Sangli District also recorded a large increase in the number of registered moneylenders, from 183 in 1965 to 250 in 1974. This is despite the gradual build up of formal financial institutions in the rural areas, under the national policy of bringing the banking culture to the countryside. But from 1974 onwards, the number of registered moneylenders took a dramatic downturn. This followed in the wake of the Debt Relief Act that was passed in Maharashtra State in 1976. The Act partly imposed a liquidation and partly a moratorium on debts of the have-not and have-little debtors, and a return of their pledged properties—even forcibly by the police if need be.

Although many debtors remained reluctant to take action against their creditors, some pawnbrokers suffered considerable financial losses. This Act was the last in a series of highly antagonistic legislation in the State between 1939 and 1975.[27] Weary of the continuous harassment caused by Moneylenders Acts and Debt Relief Acts, some lenders finally folded up and closed shop. Others simply went underground to continue without a license or restricted activities to lending only to other moneylenders. Both Hill (219) and Timberg and Aiyar (298) refer to similar laws in other States, whose onerous regulations have caused professional moneylenders to discontinue lending to agriculturists.

The Act of 1975 also unwittingly caused an increase of competition from those co-operative-type institutions that formerly had little or no interest in pawnbroking. One of the main effects of the Debt Relief legislation was that the traditional sources of credit dried up for the most vulnerable classes (Seven Decades: 435). This made the RBI send a circular in 1976, urging the District Central Co-operative Banks to sanction consumption loans against the pledge of gold and silver ornaments, a field into which these banks had seldom before ventured (compare with page 83).

In 1985, only 95 moneylenders had themselves registered in the District*, down from 250 ten years earlier. Like the other lending institutions—the banks, the Urban Credit Societies and the bishis—the great majority (84 per cent) are located in the irrigated areas.

Whether the overall scale of private lending has decreased is hard to tell. In a relative sense this is surely the case. A number of the old and established names have withdrawn from the field, partly because of the legislative crusade against them, partly because of increased competition from formal financial institutions. One of the largest jewellers of Sangli town, who claimed to have retired from moneylending, used to handle some 3000 small loans annually[28]. Others withdrew because they found better

---

* A license must be renewed every year and costs Rs 100.

investment opportunities for their money elsewhere (cf. Franda: 141).

On the other hand, a new class of moneylenders has emerged, a fact that is reported by several observers elsewhere in the country (Harriss, 1981: 166; Kurup: 1002). The newcomers consist of petty civil servants, teachers and clerks. They are the people with a regular income and access to credit from banks and co-operative societies. In Sangli the phenomenon was very marked in the small industrial town of Ramanandnagar (pop. 8000) where a number of the Kirloskar factory workers practised lending after factory hours. Also a growing number of farmers seem to be taking institutional loans for onlending to others. Playing it safe, they advance money against pawned sureties. Of course, the newcomers are part-timers only, operating without a license and with a limited number of customers. Small-scale pawnbroking appears to have become a booming sideline in India. In a personal communication to the author, Nayar, who has done extensive financial research (1973, 1982, 1986) commented in 1981 that he noticed a mushrooming of pawnshops in Kerala and Tamil Nadu.

Bastiaansen (34) notes that the typical registered moneylender of Sangli resides in the town and combines lending with the gold trade, goldsmithing or running a jewellery shop. He is, however, not a mere pawnbroker; he also mortgages land, houses, vehicles and grants personal loans on the reputation of the borrower. He is a professional. Here, only his pawnbroking activities are considered.

Licensed urban houses lend against gold only; very few still accept silver. Their counterparts in the village are more flexible and also accept silver, brass, or even cloth. The typical urban pawnbroker is a male; socially he belongs to the middle or upper class. Unlike his unlicensed counterpart, he is normally a respected member of the community, certainly now that a great many locals have replaced the original immigrants. The profession is not restricted to any particular caste.

Private lenders do their own appraising. They generally give larger advances than commercial banks, up to 90 per cent of the gold value of the pawn in extreme cases; the average, however, is close to 75 per cent. Principally, the percentage financed depends on the lender-borrower relationship. A new applicant starts with a relatively lower advance than a well-known and trusted customer with a long borrowing record.

The average broker serves a kaleidoscopic mixture of patrons at the lower income levels: petty clerks and petty traders, factory workers, government employees, but also out-of-town farmers and daily labourers. Some lend exclusively to big businessmen who need short-term working capital, or to fellow pawnbrokers who have come to repledge with their richer colleagues. In both cases loan amounts can run into tens of thousand of rupees; moreover, funds are often needed on very short notice, with a minimum of fuss and paperwork. For this type of credit the licensed moneylender is preferred to the banks.

The lender-borrower relationship is one of trust. A new customer is seldom served without a recommendation, an acknowledgement of reputation, and proof that the proffered articles are clean, that is, not stolen. Mediating on behalf of a borrower for a higher advance or to get him a loan at all, is still common practice. Secretaries of a village PACS are known to have introduced members of their society to an urban moneylender. When the relationship between borrower and lender has become a durable one after a number of years, its nature may change; the borrower no longer needs to deposit a valuable as security—the pawn loan has become a personal loan.

In earlier years, the urban moneylender had a number of agents working for him in the surrounding countryside. Agents worked on a commission basis and vouched for their client whom they accompanied to the office of their boss. The system is still in place, although in a somewhat different form. The village agent has become a petty moneylender himself, who from time to time repledges with the urban pawnbroker.

Two sets of interest rates apply to private pawnbroking. One is prescribed by the RBI and is unrealistic. Under this set of rates, agriculturists, for example, are to enjoy a concessionary rate of 9 per cent per year. This must be compared to the 17 per cent that the moneylender himself has to pay when repledging with the bank, or the still higher rates when replenishing his funds elsewhere. Not being inclined to philanthropy, a moneylender will not refer in his accounts to his landholding clients as 'farmers' but, as 'ox-cart owners' who may then be charged the maximum rate allowed by the RBI, which is 18 per cent and which is also noted in the loan contract. This often is still not the effective rate. Many pawnbrokers charge 2 to 3 per cent per month for a gold advance, depending on the loan amount, its purpose and the relationship with the borrower. Loans for a very short term may carry an even higher charge. These rates are consistent with those observed by others in India (Timberg and Aiyar: 280; Kurup: 1005; Sivakumar: 848; Harriss, 1981: 166). Of course, these rates do not appear in the written contract or in the accounts of the moneylender. But private brokers, unlike some banks, do not charge compound interest rates.

A gold advance from a registered moneylender therefore carries a higher interest rate than the 16 to 18 per cent a year levied by formal financial institutions. But the interest rate alone does not determine the choice between the two. Borrowers also take into account the extra transaction costs and the measure of inconvenience in obtaining a loan.

Compared to the impressive paperwork and the time spent in getting a bank loan, a visit to the private moneylender comes almost as a treat. His shop is open most of the day and in the evening. Appraisal, administration and handing over the cash take about 10 minutes. The pawnbroker writes out a contract form, a copy of which is meant for the office of the Deputy Registrar of Co-operative Societies. The loan amount is 'clean,' that is, no deductions are made for valuation, forms, insurance or whatever extra charges the formal institutions care to make. Nor is interest deducted from the loan beforehand, as is customary in some other countries. Transaction costs are thus kept to a minimum.

The moneylenders whom Bastiaansen interviewed claimed to handle between 100 and 500 pawn loans a year, a figure that is probably underreported. The average loan sum appears to vary between Rs 500 and 1500. Extremes, however, do occur at both ends of the scale. Lenders who deal only in large sums serve fewer customers but lend average amounts of Rs 10,000; in contrast is the jeweller who handles gold advances below Rs 50 for the sake of goodwill and the promotion of his jewellery shop.

The private pawnbroker is usually well informed about the purposes for which a loan is sought. Customers tend to volunteer information spontaneously, as if withholding it might work to their disadvantage by creating an impression of great financial need and thus lead to a higher rate of interest. Pawnbrokers are known to charge higher rates for consumption loans than for production loans. From experience they know that a production loan, for example, to buy cattle feed or fertilizer or goods for petty trade, stands a greater chance of early redemption. The loan they try to avoid is the one to a borrower with a reputation for gambling or drinking.

The use pattern of gold advances from private lenders is usually identical to that of institutional advances. The majority are for consumption, weddings and ceremonial expenses, with peaks during the marriage season; the rest follow the seasonal crop cycle while a small number of substantial loans are made to big traders and other moneylenders. In the wake of the recent dairy boom, loans are also given for the purchase of milch cattle.

The pattern of loan terms in the private sector is also not very different from that in the institutional sector. As a general rule, the smaller the loan the longer the repayment period. Because private gold advances are less substantial than institutional ones, they also take longer to be repaid. The largest loans are redeemed earliest, and many pawnbrokers therefore charge a minimum interest period of one month or fifteen days. The smaller loans may run for one year and longer.

Like all moneylenders, the private pawnbroker prefers liquid funds to frozen funds and therefore dislikes loan default, even when the proffered collateral is sufficient to cover the debt. He will try to extend the loan term rather than to auction or sell pawned articles. Public auctions require the official permission of the Registrar and involve red tape. Moreover, they are not appreciated by borrowers who see pawnbroking as a matter for discretion rather than publicity. Lender and borrower will therefore try to arrange a settlement when loan extension is no longer possible. Such a settlement often comes in the form of a quiet, private sale of the pawns.

9.2.2. *The unlicensed pawnbroker*

Of course, lending and borrowing are and always have been part and parcel of daily life in rural India. People tend to build around themselves protective networks of security, stretching from relatives and friends to patron and landlord, trader, shopkeeper, priest, fellow artisan. Such networks are necessary for sheer survival. In the context of these networks much community-based credit has social undertones and is generated on a basis of reciprocity. Loans from the unlicensed pawnbroker, however, are granted on a commercial basis and seem to carry few social implications.

Four of Sangli's eight subdistricts are located in drought-prone semi-arid zones. Farming is seasonal and extremely risky, and the need for credit is particularly pronounced. With little institutional credit available, much of this need has traditionally been satisfied by the locally based village moneylender operating without a license.

Contrary to the mutual savings and credit associations of the informal sector, these unregistered moneylenders are not easily accessible to the researcher. 'Much secrecy usually surrounds village credit-granting, which, in the absence of binding documents, is a species of personal relationship which both parties are loath to discuss' (Hill: 219). Data on operators and their procedures, number and volume of transactions are thus difficult to verify and tend to become speculative. What follows is best described as a summary of impressions and educated guesses, compiled from interviews with lenders, borrowers and interested observers of financial transactions in Sangli.

One such educated guess is that five to ten informal moneylenders operate in each village with a population of at least 2,500. Unlike his licensed counterpart in the town, the rural pawnbroker operates illegally, has neither a jewellery shop nor a goldsmithy, is not a full-time professional moneylender, can be female or male, and as a moneylender is held in low esteem by the community (Bastiaansen: 47–8).

A great many of these informal lenders are only petty pawnbrokers who lend money occasionally to make a profit on the side. Their numbers, unlike those recorded for registered moneylenders, appear to grow rather than diminish. This is probably related to the sharp rise of the world gold price in the 1970s. This greatly increased the popularity of gold among savers, particularly in India where the habit to pay dowries in gold has a long tradition and where the rise in the price of gold has been spectacular. When the introduction of irrigated cultivation and the availability of improved agricultural technology caused the demand for loans to rise, it was almost inevitable that part of this hoard of gold that had increased so much in value would be pledged as security for loans. The greater opportunities for a rather safe way of moneylending attracted new entrants into the market. In the same way that a boom in consumer demand for certain commodities boosts the number of petty traders in those commodities, so the increased demand for loans attracted a new species of petty moneylenders to the business of pawnbroking.

Harris (1981: 166) and Kurup (1002) already reported for Tamil Nadu the emergence of a new class of informal moneylender, such as the minor government official, petty clerk and teacher. This new class appears also in Sangli District, but there are others, too. For instance, in Ramanandnagar, an agro-industrial town in the centre of the sugarbelt, roughly 25 factory hands operate as part-time lenders; elsewhere the new moneylenders are the labourers of the sugar factories. They enjoy the comforts of a regular salary, access to banks and co-operative credit and, sometimes, the funds of other lenders. In addition, there are teashop owners and petty traders who have over the years managed to scrape together a bit of capital and now operate as free lance financiers. A number of these are female. Finally, former agents of the licensed urban moneylender have sometimes started out for themselves on a small scale. Most of these petty pawnbrokers belong to low income groups.[29]

Unlicensed pawnbrokers serve a clientele at the poor end of the spectrum, which also includes female borrowers. The poor, as usual, have only limited access to co-operative and bank credit and are also kept at bay by the licensed moneylender because of the unattractive smallness of their loan demands. Each village pawnbroker probably serves no more than ten customers, who may, however, borrow more than once per year. The number of clients is small, because the pawnbroker himself has such limited funds. Another reason is the sensitive nature of the relationship between lender and borrower.

The confidential nature of the financial transaction also prohibits unlicensed lenders from operating outside community boundaries, unlike the licensed urban moneylender who does serve a number of debtors from outlying villages. A lender must be completely sure of the reliability of his borrower. After all, the transaction between the two is illegal and the law of 1975 still gives the debtor the possibility of registering a complaint and getting the police to confiscate and return the pledged articles to him.

The loans of the petty pawnbroker are usually small, from Rs 100 to Rs 300. A few are taken for production purposes and may be more substantial; a recent report mentioned loans for the purchase of a buffalo, to be repaid from the proceeds of milk sales (Kortenhorst and Zevenbergen, personal communication, 1988). But it appears reasonable to assume that most are to cover consumption needs which have become more and more pronounced since Sangli has suffered three consecutive years of drought that have caused wells to run dry and crops to fail. The introduction of sugar-cane in the district also brought a new type of customer, the seasonal cane cutter. Reputedly, they visit the pawnbroker to draw loans for drinking and gambling, but this information could not be confirmed from visits to cane cutter camps.

Informal pawnbroker credit may also serve the purpose of protecting a person's credit rating in the community. Having taken a loan from a friendly source, which falls due at an inopportune moment, the embarrassed borrower eventually repays with the proceeds of a loan obtained from the petty pawnbroker. A variation on this theme is the taking of a very small loan without really needing one. By paying back promptly one tries to improve one's credit rating with the moneylender, with the ultimate aim of applying for a larger loan in the future.

The real attraction of the petty pawnbroker to the low income borrower is his willingness to accept silverware, copper and brass pots, or even second-hand clothes; the broker may also accept pawns of little commercial, but highly symbolic or emotional value such as the *mangala sutra**. In addition, he may advance up to 90 per cent of the estimated value of the pawns. The village pawnbroker is not a trained and experienced appraiser; his loan rates are not standardized along a set scale of market value, but depend rather on the reliability and trustworthiness of his debtor. New borrowers will be offered bottom value while some of their pawns may be rejected.

---

* The *mangala sutra* is a gold plated necklace symbolizing the marriage bond, like the wedding ring in Western culture (Bastiaansen: 20).

Loans from village pawnbrokers are usually short-term and repaid within two to four months. Both parties are interested in rapid redemption, the borrower because of the high interest, the lender because he fears that the pledged articles (those not of gold) might decrease in market value or be difficult to sell. Many moneylenders will charge a minimum of fifteen days, or even one month's interest.

The petty village moneylender is in the pawnbroking business for the money, and extracts a high price for his services. The interest rates that prevail in this sector of the informal financial market commonly range from 5 to 10 per cent a month; extremely short loans may carry interest at one per cent daily. Rates tend to vary with the market value and market sensitivity of the pawn, the amount of the loan and the relationship between lender and borrower. Anything outside the usual pattern, such as a loan of less than Rs 50 or one of extremely short or, strangely enough, extremely long duration(which will cause anxiety about the marketing prospects), is penalized with a higher rate. It is rumored that some lenders will take advantage of a borrower who appears to need a loan both badly and quickly by charging a higher interest rate.

Undoubtedly it is this apparently usurious rate of interest that has caused the village lender's bad reputation and low social standing in the community. But these rates are not totally without reason. Loans are small and short term; pawns must be stored and guarded; the pawnbroker may run out of funds and may have to borrow or repledge himself, when a customer needs a loan urgently. Further, goods like the mangala sutra, and also copper and brass vessels, have limited market value; accepting these as collateral imposes a risk that few lenders are prepared to take, particularly when borrowers have a reputation for drinking and gambling. Finally, the unlicensed broker operates illegally and may face punishment when caught.

Together, these factors make petty pawnbroking sensitive to cost and risk, which should be considered in any discussion of lenders' behaviour. A high rate of interest is, by itself, insufficient proof of monopolistic exploitation and excessive profiteering. As Harriss concluded from her research in North Arcot, the necessary conditions for monopoly are lacking 'small number of participants, protective barriers to entry or control over substitutes for money' (1981: 166). In the end, the very stigma of usury itself may induce irate lenders to raise the interest rate by a certain 'annoyance' factor. The stigma thus works counterproductively by boomeranging back on borrowers.

9.3. PAWNSHOP ECONOMY

9.3.1. *The licensed pawnbroker*

Reliable data on pawnshop economy and profitability cannot be presented for lack of in-depth studies of private pawnbrokers. The calculations given below should therefore be seen as an exercise in approximation of the economy of an average and only moderately popular licensed pawnshop that operates under the following assumptions (Bastiaansen: 42).

The number of advances is 300 per year; the average loan sum is Rs 3000 for the loans of a duration of three months and Rs 1000 for all others. Redemption occurs according to the following schedule:
15 per cent of the loans are repaid in three months
40 per cent of the loans are repaid in six months
35 per cent of the loans are repaid in one year
10 per cent of the loans are repaid in two years.
The interest rate is 24 per cent per year or 2 per cent per month.
The owner is the sole operator and has no hired help.
These assumptions lead to:
(A) Annual turnover of loans Rs 390,000[30]
(B) Average outstanding loan amount Rs 258,750[31]
(C) Required working capital 110 per cent × (B) = Rs 284,625
    (assuming 10 per cent idle capital)
(D) Annual interest received 0.24 × (B) = Rs 62,100
This interest income is reduced by annual costs estimated at:

moneylending license	Rs 100
inspection charges[32]	Rs 500
insurance	Rs 2000
safe locker[33]	Rs 250
shop rent, imputed	Rs 1000
tax, representation[34]	Rs 1150
sundries (electricity, stationary, postage)	Rs 1000
(E) Total annual costs	Rs 6000

The costs of using borrowed funds—which would reduce the lender's profit—are not computed in this example. The cost of labour is also excluded. On the other hand, the actual average lending rate is probably higher than the 2 per cent a month calculated here.

This results in an annual return on capital of (D)-(E): (C) or Rs 62,100-Rs 6000 : Rs 284,625 which is about 20 per cent. This is not a very high return, considering that one can earn 12 per cent on a fixed deposit with a bank or UCS. The 8 per cent difference is a small enough reward for management and risk. The moneylending angle, however, represents only part of the business of the pawnbroker, who also makes his profit from trading in gold and other valuables.

Of course, these figures are highly speculative. For instance, the annual turnover of 300 loans represents undoubtedly underreported averages. But then again, expenses may also be underestimated, considering that they are only one third of the expenses calculated for private pawnbroking in Sri Lanka in 1981 (Bouman and Houtman). Everything considered, the eventual outcome of a 20 per cent return on capital from the lending business only, might not be too wide off the mark.

9.3.2. *The unlicensed pawnbroker*

The capital of most unlicensed pawnbrokers is too small to permit more than a few loans per year. There would be no point in estimating a pawnshop economy on the basis of so few transactions. Illegal moneylending is a sideline that attracts several participants in every village. Despite its risks, the profits are apparently sufficiently rewarding to entice new entrants into the market. The illegal and private nature of the loan transaction may raise its costs and give rise to some monopoly elements; at the same time, the existence of five to ten petty pawnbrokers per village provides a measure of competition that is sufficient to prevent excesses.

9.3.3. *Institutional pawnbroking*

Individual case studies of institutional pawnbroking are not available. The pattern of gold advances in the formal sector of the financial market is much too erratic to allow even a generalized approximation of pawnshop economy. One institution may specialize in, and the other look down on, this mode of lending. Gold advances by the DCCB's head office in Sangli town approximated 5 per cent of·its loans in 1985 (but still represented a volume of Rs 26 million!); those of the Urban Co-operative Bank in Vita amounted to 27 per cent. The Sangli Urban Co-operative Bank sanctioned 1350 pawn loans in that year, as compared with some nationalized commercial banks that authorized only a handful. Although most UCS do not even have a provision in their by-laws allowing them to grant gold advances, at Ambika UCS in Wangi 40 per cent of loans are based on gold pledging. The respective differences in loan volume are also great, depending on whether the institution reckons among its clients private moneylenders who seek its money to refinance their own lending business.

It must be clear, however, that the handling of gold advances is much more remunerative than any other form of lending because of lower transaction costs and risk. While overdues on agricultural loans and other priority sector credit by banks, PACS and UCS reached epidemic proportions in all of Maharashtra, non-redemption of a gold loan is rare. Yet many institutions, the RBI apparently among them, appear to frown on gold advances.

In this, India is not alone. Chandavarkar, citing McLeod, expresses his surprise at the identical situation in Indonesia, while Bouman and Houtman do so with regard to Sri Lanka (Chandavarkar: 5; Bouman & Houtman).

The rural economy of most Asian countries can well be characterized as a penny economy, in which transactions between households and between them and rural firms are frequent and of extremely low volume. 'Viable financial intermediation between participants in a penny economy needs a mode of operation that is adapted to the nature, size and frequency of transactions in such an economy'. (Bouman & Houtman: 72). In terms of banking practice this implies low overheads, low risk and quick procurement of loans in order to attain a sufficiently high volume of business. Pawnbroking of gold and valuables meets all three requirements. It remains one of the mysteries of banking wisdom in Southeast Asia as to why those financial institutions that have been given the express tasks of promoting rural welfare and of providing the poor with access to institutional credit, seem so diligently to avoid a profit-making enterprise that enjoys such great popularity in the private sector and has so much appeal to savers and borrowers.

9.4. SUMMARY

The growing supply of institutional credit since the mid-sixties has brought many villagers into the fold of banks and formal co-operative societies without making them relinquish their links with the informal credit market. One such link is with the private pawnbroker/moneylender. Most pawnbroking in Sangli District is based on gold; this has always been a popular form of lending and borrowing because of its low transaction costs and risk. Social reformers often portray pawnhouse credit as an expression of poverty; it would be more accurate to explain it in terms of the continuing popularity of gold to savers.

The institutional financial sector in Sangli also offers pawnbroking facilities. Commercial banks serve a few low-income clients but provide medium-sized loans of Rs 1000–3000 to farmers and businessmen of some substance as well as large sums of Rs 10,000 and above to private moneylenders who repledge the valuables of their own clients. Co-operative banks and UCS, some of which make it a policy to specialize in gold advances, serve customers of lesser means. All institutions charge from 16 to 18 per cent annual interest which, plus transaction charges, push the total cost of borrowing to over 20 per cent. Loan application procedures in this sector often cause some inconvenience to borrowers. Three-quarters of these loans are repaid within half a year. Default is minimal and settled unofficially.

The private pawnbroker/moneylender appears in two forms: the urban registered moneylender who has obtained an official license and the ubiquitous village petty pawnbroker who operates illegally without a license.

Between 1950 and 1985 the number of licensed moneylenders has varied in cadence with the efforts of Maharashtra's legislators to wipe them out. Continued harassment finally forced many private lenders to forsake their license, cease lending or continue underground. The number of licensed moneylenders in 1985 was 97 as compared with 250 in 1974. All these moneylenders combine gold advances with the business of refining and trading in gold or operating a jewellery shop. The bulk of their customers come from the middle-income classes of farmers, traders and the white and blue collar professions.

The licensed pawnbroker charges an interest rate of 1½ to 3 per cent a month, but no other costs. The official Indian policy of promoting rural bank branch extensions and making life difficult for the informal financial sector has caused moneylenders to loose clients to the institutional sector. Yet, despite their slightly higher interest rates, many borrowers still prefer the licensed pawnbroker because of greater accessibility and flexibility and superior service.

Both the formal financial institutions and the licensed moneylender are concentrated in the irrigated sugarbelt. It is estimated that individual outlets in each sector make approximately 300–400 loans per year. The majority of these loans are for consumption, weddings and ceremonial expenses, but a sizeable number, coinciding with the monsoon cropping season, are to cover cultivation expenses.

The illegal petty pawnbroker differs from the licensed one in practically every respect. Lenders can be either male or female; they live in villages in both dry and irrigated zones of the district. They are not professionals and have no goldsmithy or jewellery shop, but are small-time operators who profess moneylending as a sideline. Their number appears to grow rather than diminish, due to the emergence of a new class of lenders who have the advantage of a regular salary and access to institutional credit sources. They accept other pawns besides gold objects, cater to the poor and low-income groups and serve only a handful of customers, including women. They run a risk of default and financial loss, for which they charge high interest rates of 3 to 5 per cent, in extreme cases even 10 per cent a month. Yet, there is no question of monopolistic exploitation. Five to ten lenders operate in each village (with a population of 2500), and entry into the market is free.

More than anything else, developments in pawnbroking seem to confirm the growing popularity of saving in gold, which is subsequently pledged for a loan. Now that procedures in the bishi, the standard informal savings and credit association of the district, are becoming more formalized, chances are that pawnbroking will become a more common way of borrowing and lending.

Formal financial institutions in India are charged with promoting rural development by paying particular attention to the less affluent, who are engaged in small-scale operations and enterprises. One way to do this is via a pawnbroking service. Gold pledging is not only a popular method of borrowing within this group; it is also a way of advancing money at low risk and transaction cost and must therefore appeal to lenders. The banking community in India, but also elsewhere in Asia, has apparently not yet fully grasped the great potential of pawnbroking for bringing this target group within their reach. Ironically, bankers refer the small borrower to agents of the informal financial market, the very market that the government is trying to neutralize through onerous legislation.

CHAPTER 10

Dairy development and informal moneylending*

The dairy industry in Sangli developed partly as a result of the cultivation of sugar-cane, for excellent fodder can be made from sugar-cane tops, stalks and molasses. Spread throughout the sugarbelt are many small processing plants that produce the high quality concentrates that the crossbred cows require. The dairy industry generated a milk boom on top of the sugar boom which was doubly welcome in Sangli's rural economy because, when it came in 1975, sugar was already loosing its momentum.

The milk boom merits special attention because of the rise of a new type of informal lender, the milk collector; and because moneylending developed primarily from efforts by milk collectors to expand their business and market share, and not as a result of credit demand.

10.1. THE PRIVATE SECTOR

Until recently cattle were kept in Sangli for subsistence and traction rather than for commercial reasons. The keeping of cattle for meat production is unusual in vegetarian Hindu India. Milk from local cows and buffaloes, both producing a few litres per day, was largely consumed by the farm family. There was no local milk market of any significance and the district lacked the necessary infrastructure to produce for the Bombay market.

The first push towards an infant dairy industry came from the private sector. In 1948 Chitali, an enterprising farmer and milk merchant with his own herd of buffaloes, started a small pasteurizing plant in Bhilawadi, Tasgaon subdistrict, to sell milk

---

* Data for this section were collected by H.van den Bogaard (1986).

locally and in Pune, 250 kilometres away by railway.

Initial progress was slow. In 1954 the plant processed an average of about 2000 litres of milk per day. In 1966, after electrification had arrived, output increased to 10,000 litres and again to 30,000 litres per day in 1975. In 1986 Chitali's factory processed an average of about 60,000 litres daily, collected from more than 6000 farmers in 400 different localities. More than half of this output is transported by his own fleet of refrigerated vans to Pune and sold. The Chitali family have their own distribution points in Pune, from where milk is delivered in small polythene packs by an army of cyclists and street vendors to consumers, milkshops and teahouses. The Chitalis also own several dairy product retail shops in Pune, a city of almost two million inhabitants (FAO: 101).

The Chitali businesses are very well organized. In Pune as well as in the district's catchment area, a very efficient system of collection, transport and distribution guarantees that fresh milk is brought from farmers to consumers within twelve hours, and pasteurized milk within twenty-four hours. The Chitalis also provide free veterinary service, loan guarantees for bank credit and deliver fodder to farmers' doorsteps. Any surplus milk in the flush season (September through February) is made into other dairy products and sweets. The family also takes an interest in social and community work (FAO: 103–4).

The Chitalis are, however, not without competition, both from the private sector and from State and co-operative enterprises. Their main rival in the private sector is Thote, who started his dairy plant in Ashta, Walwa subdistrict, in 1955. He also sells buffalo milk on the Pune market and his plant now produces roughly 30,000 litres per day (1986). His business is organized much on the same lines as that of Chitali. Besides these two large entrepreneurs, a dozen or so other smaller milk merchants operate in the district, selling in the urban centres of Sangli through their own milkshops or street vendors. Milk consumption has without doubt increased a lot because of the sugar industry bringing so much money and employment to the district.

The private sector deals only with buffalo milk, and not with cow milk. Buffalo milk has a higher fat content and is sweeter than cow milk. It enjoys consumers' preference and fetches a higher market price; in 1986 this price ranged from Rs 3.36 in the flush season to Rs 3.90 in the lean season (March through August) against approximately Rs 3 for one litre of cow milk. It is estimated there were about 135,000 female buffaloes in the district in 1986 (van den Bogaard: 42); the daily milk production for the market is estimated at 150,000 litres (Ibid: 44). This represents annual revenues of approximately Rs 200 million to cattle owners, not counting milk that is consumed on the farm or sold or exchanged in the neighbourhood in small quantities.

Even more important is the regularity of this income to cattle owners. The local variety of buffalo can produce from two to six litres milk per day in the flush season. Yields are lower in the lean season, especially in the arid parts of the district, when they come down to only a trickle. But even one litre of buffalo milk a day represents an income of over Rs 100 a month, with the promise of more in the flush season.

The female local buffalo has become the favoured animal of the small farmer and daily labourer. These low-income earners have neither the means, facilities, or risk-bearing capacity to dare invest in the imported and expensive crossbred cows that have a production potential of 20 litres of milk per day. Significantly, the market in cow milk has been captured mainly by large farmers with the assistance of village co-operative societies and the State-owned dairy factory at Miraj, the second largest city of the district. The co-operative societies operate in the market for cow milk without the competition of the private sector. The latter deals solely with buffalo milk and in that market it competes fiercely with the co-operatives for the goodwill and business of the farmers.

10.2. THE STATE AND THE CO-OPERATIVE SECTOR

Until the mid 1970s the State of Maharashtra still relied on milk from the neighbouring State of Gujarat and milk powder from the European Economic Community to meet the demands of consumers in the Greater Bombay area. This prompted the State government to follow the example of Gujarat and start its own dairy development programme. But unlike Gujarat, that promoted buffalo milk and let market forces decide the production and consumption price, the Maharashtra government opted to promote high-yielding crossbred cows. The milk production from cows is much less subject to seasonal fluctuation than that of buffaloes.

In Sangli District the programme started in 1975. To stimulate farmers to adopt crossbred cows, they were offered cheap loans and occasional subsidies of up to 25 per cent of the purchase price of cattle. To encourage them still more, the government dairy factory at Miraj guaranteed to pay farmers a cow milk price equal to that of the higher quality buffalo milk that normally fetched a market price of about half a rupee more per litre.

To get the milk from the farm to the factory, the government stimulated the establishment of a co-operative dairy society in every village. This policy was pursued with much (political) vigour and speed, regardless of the viability of each society. While the number of milk societies in 1970 was only 30, this had grown to 128 in 1975 and to 459 in 1985; the total number of villages in the district is 546 (Socio Economic Review, 1980–1).

In Miraj the milk is pasteurized and sent to one of the three State-owned dairy factories in Bombay that make up the Greater Bombay Milk Scheme which distributes milk and dairy products to the public. The co-operative milk sector is organized less efficiently than the private sector when it comes to collection, processing and distribution of the milk. It takes the latter only one day to bring the pasteurized milk from producer to consumers in Pune, against three days for the co-operative sector (van den Bogaard: 45).

The government efforts to stimulate dairy development proved remarkably effective. Until 1975, milk receipts at the Miraj factory consisted predominantly of buffalo milk. But from that time, due to the attractive milk price and superior production of the crossbreds, the processing of cow milk at the plant increased very fast and soon surpassed that of buffalo milk. In 1981, 75 per cent of the average daily procurement of 120,000 litres consisted of cow milk. Within six years, the sale of cow milk to the dairy factory in Miraj had surged from a few thousand litres per day to a daily average of 90,000 litres.

But soon this growing flood of milk began to subside. In effect, government measures proved so effective that they overshot their mark. In 1981 the Bombay factories were flooded with milk supplies and had to cut back acceptance drastically, leaving the Miraj factory to deal with the oversupply itself. This was done by converting fresh milk into powder and storing it *in situ*. Meanwhile the factory still had to accept all the milk offered and pay the attractive minimum guaranteed price. When its storage capacity was exhausted, the factory was forced to destroy part of its milk in 1985. Within ten years' time Maharashtra's market for milk was saturated, at least against the retail price of Rs 5 to Rs 6 per litre to the consumer in that year.* This is a remarkable performance and it has benefited many cattle-owning farmers; but it also forced the government to review its dairy policy and to request the national government to stop imports of EEC milk powder and help it export its own surplus of milk and dairy products to Delhi and other big cities.

Meanwhile the Maharashtra policymakers turned their attention to buffalo milk. In 1981 the former price relation between cow milk and buffalo milk was restored by guaranteed prices based on the fat content of the milk (which also discouraged dilution of milk with water). While the guaranteed price of cow milk was between Rs 2.77 and Rs 3.13 per litre, the price of the much richer buffalo milk was between Rs 3.46 and Rs 3.92 a litre (flush season versus lean season).

The government also promoted the adoption of an improved breed of Gujarat buffalo. This new breed could produce from 8 to 15 litres of milk a day, against the 2 to 6 litres of the common local variety. Its purchase price, however, was also much higher, Rs 5000 to Rs 6500 compared to Rs 1500 to 3000 for the local type. The government, therefore, introduced a scheme of subsidized credit on favourable terms and a guarantee by the milk co-operative society to secure repayment of the loan through sales of milk to the factory. The measures caused an increase in the intake of buffalo milk at the Miraj factory after 1981. The intake rose from 33,000 litres to 45,000 litres per day in 1985.

Total average daily milk sales by Sangli farmers* to the Miraj factory and handled by co-operative societies, were estimated at

* Six Indian rupees equalled approximately US $0.50 at January 1988.
* Milk sales by farmers from neighbouring districts, who also send their milk to Miraj, are not included in this figure.

115,000 litres in 1985 (van den Bogaard: 44). Together with the estimated 150,000 litres of buffalo milk handled by the private sector, this constituted an average daily milk flow of 265,000 litres, representing an income of more than Rs 300 million per year, not counting private sales and home consumption by farmers and other buffalo owners[35].

On top of this sum, which is quite a substantial one for a district of 300,000 households, is the money that is earned by an army of several thousand milk collectors and distributors who bring the milk from the farm to the consumer. The dairy plants, the small cattle-feed producing units, the many milk shops and street-vending businesses provide additional employment.[36] Milk production, processing, distribution and related support activities make the dairy industry as important as the sugar industry in the economy of the district.

The buffalo is the small farmer's favourite animal. Although the superior milk production of the improved Gujarat species is still less then the 10 to 20 litres per day that a crossbred cow can produce, the buffalo is more heat and disease resistant than the cow, that has to be kept, fed, watered and milked in a stable. The cow is usually insured; insurance is obligatory in formal cattle-loan programmes. The crossbred, moreover, is also fed, in part, on the high quality concentrates bought from the processing plants. Investment in such cattle is both risky and expensive for resource-poor farmers. The earlier dairy development programmes and subsidies of the government to stimulate cow-farming, therefore largely bypassed the small farmer. But it was different with the buffalo-promoting programme, that had great appeal for small farmers. Here, the private sector also assisted them to benefit from the programme through loan guarantees and a direct supply of credit.

10.3. MILK COLLECTION AND MONEYLENDING

The private commercial sector runs an ingenious and very efficient system for collection of buffalo milk. The factories of Chitali and Thote have their own networks of collection centres, numbering 40 and 25 respectively. These centres are run either by their own employees or by commission agents who receive 2 to

3 paise per litre.* A typical centre handles about 1400 litres per day during the flush season.

Each centre employs its own group of collectors to transport the milk from the farms to the centre, usually on a bicycle. Often the milk is collected twice daily.

A collector's payment is related to the fat content of the milk: the more fat, the higher his reward (minimum required fat content is 6 per cent). Buffalo milk with, for example, 7 per cent fat—which is quite usual for these areas—brings him 60 paise per litre.

Collectors thus have a vested interest in discouraging farmers from diluting the milk with water, and are present when the buffaloes are milked. This is time consuming and limits a collector's deliveries to 25 to 50 litres per day, which, however, still represents a daily income of Rs 15 to 30 or a monthly income Rs 450 to 900. These are attractive sums compared to, for example, the Rs 250 to 450 per month salary of a village co-operative society secretary, or the Rs 15 per day of a cane cutter. But in the arid regions yields of the local-type buffalo are low and the collector has to work hard for his money. In the lean season milk production may be reduced to less than 1 litre per animal per day.

To ensure that collection continues in the lean season, centres withhold, until the end of the lean season, part of the collector's money earned in the flush season. To stimulate collectors still more, centres require that collection in the lean season equals at least 50 per cent of the quantity delivered in the flush season.

The 65 centres of Chitali and Thote employ the same system of collection, stimulation and reward, providing work to more than 4000 collectors in 1988. Milk collecting is an attractive job and a cause of fierce competition between agents. There is also strong competition between Chitali and Thote, who often have centres in the same villages, and between them and the co-operative societies.

Milk collectors compete with each other for the farmer's favour and his milk by offering extra services. These services extend not only to farms in both irrigated and dry zones, but also to labour

---

* One rupee = 100 paise.

camps in the cane fields, the temporary abode of migrant cane harvesters.

There is a certain complementarity between the sugar and dairy industries and between the wet and dry zones of the district. Many migrant labourers originate from the dry zone and come to work in the cane fields with their oxen, that draw the carts used to transport cane. The migrant families bring their milk buffaloes with them; while the men work in the fields, the women tend the cattle. The cane fields supply the fodder needed by the animals and labour is partly paid in cane tops.

The milk collector's services to farms and camps consist of daily milk collection, weekly payments for milk, and supply of fodder, concentrates and credit. All these services are delivered at recipients' doorsteps at no direct charge.

Loans from the collector range from a few to several thousand rupees and are used for all kinds of purposes, even to pay the purchase price of a local variety buffalo. Most loans are interest-free. The only condition is loyalty in supplying milk to the collector providing the credit.

To maintain loan services, some collectors have started to organize their own bishi. One such bishi in Ramanandnagar consisted of twelve collector-members who each contributed a weekly sum of Rs 50. Another one in Takari, Walwa subdistrict, had grown to twenty-five members in 1986, and a weekly fund that equalled the purchase price of a local buffalo.

The private commercial sector even competes with village co-operatives for participation in the government loan programme for the purchase of high grade Gujarat buffaloes. These loans are for a maximum period of five years at a subsidized interest rate of 12 per cent. Beneficiaries must either provide land as collateral or obtain the guarantee of a dairy co-operative society to repay the loan directly from the borrower's milk delivery proceeds. The same letters of guarantee can now be submitted by the private factory owners and are accepted by the bank. The letter is given to the collector who receives the loan, buys the buffalo and contracts it out with the farmer. The collector owns the buffalo until the loan is repaid with a 12 per cent interest and a Rs 100 insurance fee.

The superior production capacity of a high grade buffalo permits repayment of the Rs 5000 loan in two years, while still leaving the farmer a handsome income. The loan for the cheaper and less productive local buffalo can be repaid within two to three years. The collector repays through milk delivery to the factory. Note that this time the loan is not given to the farmer interest-free. Still, all parties—farmer, collector and factory owner—profit from the operation.

10.4. SUMMARY

Initially, the dairy industry in Sangli developed only slowly and was restricted to the market for buffalo milk. This market was opened up by private entrepreneurs who established dairy plants in the district and sales points in distant Pune. Distribution required a very efficient system of milk collection, processing and transport. Central to the system were the collection points throughout the district, and the collectors who fetched milk from farms to these points. Through this private initiative the local buffalo, despite its low milk yield, has become the favoured animal of the small farmer, providing a regular income in an otherwise insecure existence.

Since 1975, fresh impetus has been given to the cattle industry through a government campaign of promotion of high grade cows and buffaloes that could be bought with institutional credit at a preferential rate of interest. Response was greatest among affluent farmers, who could afford to invest in the superior producing imported cows that were, however, also prone to risk and costly to care for. To obtain a loan required cumbersome application procedures, lengthy processing and high transaction costs. Those applicants with lesser means and influence who could not provide sufficient land as collateral nor come up with acceptable guarantors were greatly inconvenienced.

The informal sector has come to the aid of the small farmer. The development of private dairies has spawned a new type of informal lender: the milk collector who is the essential link between the factory and the farmer. To survive in a competitive yet remunerative milk market, each collector has to keep his milk-supplying buffalo owners happy and loyal by making himself useful.

One way is through regular lending of small and larger sums of money. It is even in the collector's interest to provide loans for the purchase of cattle to increase his market share. To this end, some collectors have formed their own bishi; they also participate in the government's preferential loan programme on behalf of farmers. This is done with the co-operation of banks that accept the guarantees of owners of private dairy plants for loans given to collectors.

Like the registered moneylender who refinances the unregistered pawnbroker through the commercial banks, the collectors have become a point of integration between informal and institutional financing.

CHAPTER 11

Evaluation and summary

11.1. FINANCE AND DEVELOPMENT IN SANGLI

India's financial system has come under increasingly tight government control through public ownership of the major financial institutions in the country. These institutions are required to direct the flow of finance in accordance with the guidelines of national and state development plans. These plans aim at eradication of poverty and emphasize assistance to priority sectors, of which the most important are agricultural and rural development and small scale enterprise.

The approach to rural development is typified by a strong emphasis on direct public investment in irrigation facilities and other infrastructure, and expansion of the formal credit system via co-operatives and banks. These are supposed to foster private agricultural investment, with the aid of preferential lending interest rates.

This approach has been followed in the Sangli District, where the government has provided the major development thrust to transform the economic environment. In the fifties and early sixties the government invested in the construction of dams and minor irrigation facilities, supported the building of co-operative sugar factories and encouraged farmers to grow cane through a favourable price policy. These initiatives have brought life into the stagnant rural economy. Further, after private entrepreneurs had started the beginning of a dairy industry, the government helped spark off a small dairy boom in the seventies by raising the price of cow milk and introducing the crossbred cow and Gujarat buffalo, both superior milk producers. Developments in the sugar industry and livestock sector have mainly taken place in the irrigated areas of the sugarbelt, but the higher incomes and employment opportunities that followed the sugar and dairy boom have resulted in an economic diversification that has benefited a much larger area.

Institutional credit and co-operative credit in particular, has been co-instrumental in this economic transformation. Small and large farmers could avail themselves of long term credit of the LDB for wells and lift irrigation, while the PACS distributed seasonal loans for the growing of sugar-cane. The DCCB provided interim credit to the sugar factories to pay for the cane supplies. Medium term credit was available for the purchase of milk cows and small livestock. When later, due to the success of irrigated agriculture, farmers became interested in other crops and activities, these same institutional sources gave loans for the establishment of orchards, betel and rose gardens and vineyards.

This process has been supported by a formidable growth of formal banking institutions, whose performance has been greatly influenced by the selective credit policies of the Indian Government. These policies call for the distribution of cheap credit to priority sectors via a great number of programmes and schemes for specific regions and target groups. Like elsewhere in India, commercial banks in Sangli have come under heavy political pressure to alleviate rural poverty by participating in these development programmes, for which they could rely on preferential refinancing and government insurance of loans. Financial institutions thus became more or less grassroots development agents involved in a great number of projects. These projects are outlined in Annual Action Plans for each district and based on norms handed down by the Reserve Bank. These Plans give highly detailed sector and scheme-wise allocations for each participating bank. Sangli's Draft Action Plan for 1985 totals Rs 422 million, covers 31,466 beneficiaries, and includes 23 different banks (Annual Action Plan: 31–2).

Although official development policies and detailed planning processes have worked well, they also had less favourable consequences. The political commitment to high volume lending has been at the expense of loan quality. The phenomenal increase in the number of loan accounts has placed a severe strain on the banks' organizational resources. It has resulted in overburdened staff, declining control, alarmingly high overdues, deteriorating customer service and low bank profitability. The recovery rates for agricultural loans of commercial banks in

Maharashtra are among the worst in India and have remained consistently below 50 per cent since the early seventies. Wealthier farmers are often the main defaulters.

Under these circumstances, banks have become reluctant to spend energy, resources and manpower on making loans outside the purview of the officially recognized priority sectors. Loan demands, such as for commerce, transport, distributive trade and consumer finance, that followed in the wake of the economic expansion, could not expect a sympathetic response.[37]

The gap between the demand for and supply of bank credit has widened, and so has the time lag between the demand for and the sanction of a loan. Requests involving small sums of money, which have never been popular with banks, arouse even less interest. This is particularly true in times of financial crises, when the monetary authorities apply a sharp brake on the expansion of credit, with undesirable consequences for certain sectors and especially for small borrowers (RBI, 1985b: 296, 308).

11.2. OPPORTUNITY AND RESPONSE OF THE FINANCIAL SECTOR: PLANNED GROWTH VERSUS ORGANIC GROWTH

These deficiencies have provoked a reaction and stimulated the activities of informal financial intermediaries. Chandavarkar compares the informal financial sector to a reactive sector 'representing a response to financial repression as well as market and organizational failures in the formal sector and meeting the demands of not only small-scale borrowers but also of the unsatisfied fringe of medium and large-scale borrowers in the organized sector' (Chandavarkar, 1985a: 26).

One has to view the proliferation of UCS and bishis in Sangli in the 1970s in this light. Improved economic conditions gave rural households the opportunity to diversify their activities and lessen their dependence on seasonal agriculture with its erratic incomes. This caused a growing demand for alternative financial institutions beyond the existing formal structure of co-operatives and banks. The spontaneous expansion and diversification of the economy that resulted from the sugar and milk boom demanded a creative approach to financial services. The conventional institutions, bogged down in rigid blueprints of financing targets, and not allowed to make their own lending decisions, were too crippled in operational efficiency to offer such a creative approach. Too much planning threatened to retard economic activity.

This accounts for the emergence and subsequent popularity of the milk collector as an informal source of credit for cattle owners; it accounts for the growing number of petty moneylenders and pawnbrokers. It also explains the growth and concentration of UCS and bishis in the commercialized irrigated areas where the changes have been greatest and the need for novel institutions more explicit. UCS and bishis are registered in such areas, despite the presence of numerous other financial institutions.

Comparative research between UCS and bishis has revealed the much greater popularity of the bishi, whose numbers have increased dramatically in the 1980s. The bishis' greatest assets are self-help, which puts savings before credit; autonomy and self-regulation; easily understandable rules and procedures that facilitate transactions and a flexible response to new demands; greater freedom from patronage and political interference. In bishis there is a constant search to adapt to changes in the socio-economic environment. Such adaptations are much less evident in the UCS.

In a sense, Sangli provides a fresh example of the familiar dilemma of organic growth versus planned growth. The district's increased money-flows and new riches are seeking outlets in investment opportunities beyond conventional agricultural activities; they have also brought higher standards of living and a demand for new consumption alternatives. These pursuits are not included in the official rural development plans and are thus ignored by the institutional lending agencies.

In its review of seven decades of co-operative banking, the Maharashtra State Co-operative Bank remarks that it is 'being saddled with mounting lendable surplus deposits seeking legitimate outlets, while at the same time rural development programmes have been starved of adequate credit. The challenge of the eighties is to tackle the complex problems that have emerged as a result of the developments of the past decade' (Seven Decades: 513). It seems that the informal financial sector has responded more readily to this challenge, employing a much wider definition of 'legitimate outlets' than the formal financial structure.

11.3 FORMAL AND INFORMAL FINANCE: BETWEEN MYTH AND REALITY

There is a rural Indian proverb that says that no village is complete without a moneylender, a medical practitioner, a teacher and a stream that does not dry up in the summer. Franda, dryly observing that the moneylender is mentioned first, is not surprised at this preoccupation with credit. In an agricultural economy exposed to the vagaries of the weather, 'moneylending has been a central part of rural life for as long as anyone has written about or talked of India' (Franda: 39).

Nonetheless, one seldom hears an Indian in a policy-making position say a kind word about the informal financial markets. India's state bureaucracies have traditionally been suspicious of informal markets and have constantly strived for their elimination. Their suspicion is echoed in the popular press and films and repeated in the works of serious scholars.

There is often a strange inconsistency in the way some Indian authors view the performance of informal financial intermediaries. A typical example is Prakash's study of private financing firms in Trichur, Kerala State. In Trichur the chitty, an informal savings and credit society of the ROSCA type, used to enjoy great popularity among the 78,000 inhabitants. This came to an end with the implementation of the Kerala Chitties Act in 1975. The objective of this Act—another example of India's efforts to eliminate, if not to control the informal credit market—was 'to regulate the chitty business in the interest of members'. Its outcome was the virtual disappearance of the chitty in Kerala because members wanted none of this official regulation

Within one year, its former organizers had come up with an alternative, the partnership firm. These firms are registered under the Indian Partnership Act and have to obtain a moneylending license. Partnerships keep their capital subscribed below Rs 100,000 to remain outside the purview of the regulations of the RBI (1985b: 337). There was an immediate interest in these firms in Trichur. In 1983 more than 1500 partnership firms were in operation, collecting savings deposits from the public and offering loans to carefully selected clients (Prakash: 2129). The economic climate of Trichur was excellent, partly because of new employment opportunities, and partly because of money remittances from Gulf countries. A sizeable amount of these remittances was spent in the town on consumer and luxury items, gold ornaments and construction materials (Prakash: 2130). This, in turn, proved a boon to retail traders and small business men and led to a phenomenal increase in commercial activities and a corresponding demand for credit. But the supply of formal credit was limited because of the official restrictions on credit for business purposes imposed on the banks since 1980. The author concludes that the credit restrictions on the one hand, the fast-expanding credit demand on the other, plus the ready availability of savings, created favourable conditions for the growth of the private financing firms (Ibid.). The situation in Trichur resembles that of Sangli once the sugar and dairy industry had begun to make their impact on the local economy.

Like the chitties before them, the partnership firms now have become a popular informal financial institution. The many savers are happy because they receive a much higher rate of interest (24 per cent per year) than available on deposits with commercial banks. Borrowers, mostly traders in need of short term capital, are also happy, although they have to pay 30 to 40 per cent interest on an annual basis (the maximum allowed by the Kerala Money Lenders Act is only 12 per cent). But access to loans is easy and without extra transaction costs (Ibid.: 2131); moreover, booming trade allows them sufficiently high profits to offset borrowing costs. Bank loans, although much cheaper, are simply not available. Finally, the organizers are happy because loans are repaid and the partnerships provide them with a decent living.

All parties to this self-help institution, therefore, are satisfied. Only the author, Prakash, is not. Although he agrees that the partnerships fill a need and are managed efficiently, he finds fault with their performance on ethical grounds. He considers it a 'serious defect that available savings in the community are diverted for trading and other such speculative type of activities' and are not used for industrial investment. The latter is apparently the author's preferred investment, although he concedes that such investment 'in Kerala is considered risky, difficult and unattractive due to many reasons' (Prakash: 2132). He deplores also the high interest rates and the fact that the partnership firms—like some registered moneylenders in Sangli—have to keep dual accounts. 'One is to satify the authorities and the other being the real account'; they thus 'generate and distribute black money' (Ibid.). He concludes that there is a need to 'regulate the activities of the firms through appropriate regulation measures to prevent a collapse of a large number of the firms in case of a future business recession' (Ibid.: 2133).

Prakash's argument seems strange as well as his view of the powers of regulation. In Trichur, the firms provide a response to formal financial repression by meeting the unsatisfied credit demands of a group of borrowers with no access to commercial banks; they protect depositors from unattractive investments in industries; and they have found a way around the unrealistic interest rate limits of the Money Lenders Act, that is satisfactory to both savers and borrowers. What more could one demand of a financial institution?

Yet, the firms' performance is rated as unsatisfactory because they are not doing what the author believes they should do. In India, policymakers and authors like to measure the performance of the informal finance sector, not on the basis of its own merits or the satisfaction of participating parties, but through the application of their own ethical and economic yardsticks.

All the standard notions and conventional characterizations of informal credit have come together in the following argument for credit reforms. 'Informal lenders exploit their borrowers by charging excessive interest rates, taking away borrowers' land, forcing borrowers to buy high priced goods, paying borrowers too little for their products and forcing borrowers to repay loans with underpriced labor' (Bhaduri, quoted by F. A. O./O. S. U.). There are many who concur with the view that 'the poor villagers are permanently trapped in the net of the moneylender because of the vicious circle of poverty: their expenses exceed their income every year so they have to go to the moneylender to borrow . . . .' (Wadhva: 179).

Such statements come easily in India but they do not explain the economics of rural lending and how the moneylender survives when his clients have less income than expenditure.

Indian friends of the author in the co-operative structure, aware of his interest in informal finance markets in Sangli, strongly denied any relationship between informal lenders and the economic development of the district. To prove their point they cited hearsay scandals of bishis, insisting that bishi-loans stimulated alcohol abuse. Financial policymakers, on the whole, routinely start from an assumption of the inherent superiority of formal financial systems in bringing about development and firmly believe that the informal financial sector tends to reduce the effectiveness of monetary regulation measures. The official Financial Review Committee Report of 1985, although granting the usefulness of some informal intermediaries as para-banking agencies[38], still wishes to 'regulate that part of their activities which are not in conformity with official credit policy' (RBI, 1985b: 94). India has, therefore, repeatedly resorted to legislation as a means to curb the activities of ROSCA and moneylenders; witness the Chit Fund Bill and the various Moneylenders and Debt Relief Acts.

Supporters of informal markets, on the other hand, are convinced that modern institutions cannot adequately serve households in rural areas, particularly low-income households. They consider control of the informal sector unfeasible. 'Attempts to regulate the informal sector are liable to fail because its very rationale derives from its informality and immunity from official regulation. Regulation is more likely to be counterproductive since it may lead to evasion of laws without providing for adequate institutional substitutes' (Chandavarkar, 1985a: 33).

The counter-productive effect is evident in the case of the Kerala Chitties Act of 1975, which was intended to regulate the State's chitty business but virtually drove them out of business instead; it is also evident in the case of the Maharashtra Debt Relief Act of 1976. This Act, rather than bringing relief to the most heavily indebted, caused traditional sources of credit to dry up for the most vulnerable classes. Eventually, it even forced a reluctant RBI to permit DCCB branches to grant pawnbroking loans to individuals on the security of gold.

One may also wonder whether laws and regulations that are clearly not in concert with actual economic realities do not, in fact, foster dishonesty rather than compliance. The unrealistic interest rate limits of the Kerala Money Lenders Act actually forced the partnership firms of Trichur to keep dual accounts. Similarly, the prescript that loans to agriculturists should carry a concessionary interest rate of 9 per cent invites evasion, when the moneylender in Sangli himself borrows his capital at 17 per cent or more when repledging with banks or others. It is not in tune with reality to expect philanthropic behaviour from professional moneylenders.

Whole volumes have been written on the merits and consequences of the so-called cheap credit policies of developing countries that force concessional interest rates on lenders (Adams; Howell; Von Pischke). Unrealistic legislation and policies seem to nourish their own culture of discontent, stimulating evasive tactics and manipulative routines, and so slowly erode lenders' and borrowers' moral behaviour.

11.4. FORMAL AND INFORMAL FINANCE: SUBSTITUTION OR COMPLEMENTARITY?

Many Indians take a good deal of pride in seeing that the ratio of formal loans to total loans has increased substantially over the past twenty-five years; they firmly believe that formal loans should and do substitute for informal loans if financial markets are developed properly (Adams, personal communication). Economists, too, suggest that in the course of development informal financiers and loan associations become redundant (Drake: 140). Remarking on the increased demand for credit in Pakistan, Khan suggests that commercial and agricultural banks should take the opportunity to replace all non-institutional sources of credit (F.A.O./O.S.U. reviewing Khan). Such substitution is also suggested by those who deplore the fact that in India only 35 per cent of rural credit requirements are met by institutional agencies.

Is such substitution really advisable and possible? And is the Indian banking system to blame for not having tried hard enough 'to redress the enormous extent to which small and marginal cultivators and other rural households continue to depend on non-institutional sources of finance', as one advocate of the formal sector suggests? (Shetty: 1427).

It is doubtful whether those who stake future rural development exclusively on rapid introduction of formal financial institutions, have actually considered the administrative and economic consequences for institutions charged with replacing all informal sources of credit through financing low income households and enterprises. Most rural economies of Asia can be characterized as penny economies, in which money transactions between participants in the economy are very frequent, small-sized and measured, as it were, in dimes and quarters rather than dollars. Viable financial intermediation between these participants requires a low-cost institution, making a sufficiently large number of profitable mini-loans to offset costs and stay in business without the support of subsidies. Private moneylenders and pawnbrokers, self-help institutions, and money clubs like bishis, are eminently suited to accommodate the financial demands of such a penny economy and survive; but are banks?

Essentially, a bank is little more than a money shop that buys money from savers and the Central Bank and sells it again to investors. It competes for business with many other money institutions. Like other shops, a bank derives its income from the margin between buying and selling. In the banking trade, fierce competition keeps this margin small and banks, therefore, need a lot of money transactions to remain viable. This they achieve most speedily by making substantial loans to large enterprises and rich customers.

Consider now the plight of an average rural bank, committed to serve, first of all, small enterprises and low income households. Give this bank a staff of four to carry out its task: two loan officers, one teller and a clerk. Consider also that it has to operate under the following conditions:

(a) Annual overhead costs are Rs 60,000.
(b) The sales margin between buying and selling is 6 per cent.
(c) Loan volume needed to cover costs = Rs 1 million.
(d) Average loan amount is Rs 1000.

(a). Overhead costs include salaries and allowances, office rent, furniture and equipment, maintenance, stationary, postage and telephone charges, sundries. To compare: the annual overheads of a typical village PACS vary between Rs 15,000 and 20,000 in Sangli District. But a PACS has a smaller and underpaid staff, it operates from humble, rentfree premises and has very little office equipment.

(b). This is a generous margin, PACS have one of close to 4 per cent. The higher margin is taken to illustrate the bank's predicament even better.

(c). $100 : 6 \times Rs\ 60{,}000 = Rs\ 1$ million.

(d). Average loans by Regional Rural Banks in India, charged specifically with serving the weaker sections of society, were Rs 835 per account in the mid-seventies (Shetty: 1443). It is here put at Rs 1000 for argument's sake. Bishi loans and pawnbroking loans in Sangli District are much smaller, averaging between Rs 100 and 500.

To achieve the lending volume of Rs 1 million that is needed to keep the rural bank alive and healthy, the two loan officers, together, would have to process one thousand loan demands of one thousand rupees each per year. A low-cost institution such as the PACS, that has to survive on a 4 per cent margin, still needs to make between 375 and 500 loans to cover its overheads. A LDB branch needs a loan turnover of Rs 4 million for survival (Five Year Plan: 49).

In contrast, *the average staff officer in the Indian banking system handles 120 deposit accounts and only 14 loan accounts* (author's italics); the highest productivity is recorded by the Syndicate Bank, a highly innovative institution, with 287 deposits and 48 loan accounts per officer (A.I.D.: 28). How, then, could the formal financial system ever expect to handle thousands of mini-accounts per institution? To make matters worse, this scenario has not taken loan default into consideration and thus paints a much too rosy picture. A loan default of 5 per cent is commonly regarded as a very good performance for a bank that operates under conditions of a developing rural economy. Even then, such a low default almost completely wipes out the 6 per cent sales margin and threatens the institution's viability. In reality, the default rate on loans by rural banks in Maharashtra has consistently run close to 50 per cent since the seventies. According to recent information from India, the Maharashtra government in 1988 voted yet again to write off a large volume of co-operative and bank loans.

Finally, most rural loans are seasonal and for less than twelve months, thus calling for a loan volume higher still than one million rupees. Doubling or tripling the margin, or even raising loan interest rates sky high, would not solve the main problem of handling the financial needs of a penny economy, which is volume. Like many other businesses, banks survive on volume and penny economies do not generate sufficient business volume.

The above, much simplified calculation should convince the advocates of replacing all non-institutional sources of credit by formal institutions, of the unlikelihood of the substitution scenario.

A bank taking over the functions of informal sector lenders would have to hire scores of additional staff and still be unable to cope. It would then become precisely the bloated bureaucracy that critics abhor, and its costs would automatically exclude any possibility of ever becoming an operationally viable banking unit. Even when financial markets in India are developed properly, the replacement of all non-institutional sources of credit by commercial and agricultural banks is simply prohibitive and utopian.

The actual process of financial market development in Sangli district is, however, much more complicated than the simple substitution scenario suggests. Not only has the supply of formal loans in the district grown substantially; but informal financial transactions have also increased rapidly with economic development, though they may have declined in relative importance.[39] Some forms of informal finance such as the mutual savings and credit societies have expanded very fast; new intermediaries, like the milk collector, have appeared; and other types, like the registered moneylender/pawnbroker, have retreated or experienced a metamorphosis. The Review Committee, in its report to the RBI on the working of the monetary system in India, concedes that 'the phenomenal growth in commercial banking business which has been in evidence since the early seventies, did not hamper the activities of non-banking financial intermediaries who expanded their business, and in a sense complemented the activities of the commercial banks' (RBI, 1985b: 91).

Both formal and informal finance have their strong and weak points. Savings deposited with a bank are secure but earn little interest; deposited with a bishi savings face the risk of loss, but also the promise of a very high return. Making large and long-term loans, based on collateral, is the strong point of formal finance; making small, short-term and unsecured loans is the informal sector's strongest point. 'Informal market intermediaries serve small scale enterprises whose size and instability make them unattractive to the commercial banks' (Timberg & Aiyar: 280). All informal lenders in Sangli encompassed in this research provided such loans.

Bishis, in particular, deal with the demands of a penny economy by sanctioning mini-sized and very short loans that cover production, consumption, socio-religious purposes, and emergency situations. According to Holst, efficient specialization and not poor performance, is the reason why informal finance agents usually lend relatively small amounts on a short-term basis and do not finance long-term fixed investment (Holst: 141). It is precisely this small-loan specialization which precludes the substitution of the informal lender with a supposedly more public-spirited formal institution.

Perhaps the greatest asset of the informal finance sector is its inventiveness and quick response to market opportunities, accommodating new categories of clients and businesses more readily than formal sector finance institutions. Innovative behaviour in Sangli is shown in bishis that change from a one-year cycle to a longer cycle to accommodate loan demands; in the change in ROSCA from pure lotteries to lotteries with discount; by UCS who require their savings' collectors to deposit a security fund to forestall fraud. It is also demonstrated by milk collectors who found in dairy plant-owners a reliable guarantee for a loan, and by pawnbrokers accepting for security, trinkets with only a high emotional value. 'By pioneering new forms of credit later adopted by the regular banking system, non-bank intermediaries form the cutting edge of banking innovation' (Timberg & Aiyar: 280).

11.5. EXPLORING NEW FRONTIERS OF FINANCIAL TECHNOLOGY

The number of formal as well as informal financial institutions in Sangli has multiplied over the past twenty years, leaving rural households greater freedom in choosing a suitable intermediary or advantageous transaction. This process has been accompanied by a constant search for the appropriate mix and the proper technology of financial intermediation.

On the one hand there is a movement away from the larger, complex and more formal organization towards the smaller, simple and informal one, because of the latter's greater social and organizational viability. Managers of the large co-operative banks have promoted the smaller UCS, and dignitaries of the UCS have championed the cause of the still smaller bishi.[40]

On the other hand, there is also the question of a reverse movement, in which bishis are converted into UCS and UCS into co-operative banks to obtain greater economic viability and access to larger financial resources to satisfy increased credit demands. Other bishis, however, that wanted to preserve full autonomy rather than losing independence by conversion into a UCS, have chosen to move from a one-year cycle toward a longer cycle that promises more durable financial services. Yet others have become so large that they have to employ a paid administrator and request pledges or guarantors as security for loans, while lengthening the term structure of their lending. Bishis have started to use printed rather than written rules and by-laws and have introduced printed passbooks and loan demand forms; book-keeping is standardized, using the printed forms that are for sale in bookshops.

Thus, on the one hand, there is suspicion of formalities and control imposed by state bureaucracies; on the other hand, bishis have started to copy some of the formalities of UCS in a search for greater security and incorporation of the good points of both worlds, the formal and the informal.[41]

The result is a veritable kaleidoscope of financial institutions catering to every taste, purse and preference, and representing a continuum between the two extremes of the formal/informal distinction. These institutions are not exclusive; cross-membership is common, and interaction between the different institutions suggests complementarity rather than simple substitution. Banks have sought actively to finance moneylenders, pawnbrokers and milk collectors; bishis deposit idle funds with banks for safekeeping; both bishi and UCS are known as interim lending agencies for clients of PACS and banks. As McLeod demonstrated for Indonesia (McLeod: 323), people in Sangli may have feet in both camps, holding simultaneous membership in bishi, PACS and UCS or having acounts with a commercial bank, whenever this is advantageous to them.

Observation of Sangli's rural financial market reveals that the popular dichotomy between formal and informal sectors should be interpreted carefully. McLeod prefers to speak, not of financial dualism, but of a multidimensional market with a range of various kinds of financing arrangements between the two extremes (McLeod: 322).

Lately a number of authors have suggested a policy of seeking actively to link informal and formal financial institutions (Ghate: 59: Holst: 145; Seibel, 1985). Seibel, through twenty years of research of indigenous African and Asian associations, has become the undisputed expert in the field. He has even designed a linkage model because of the belief that 'as long as formal and informal institutions—such as savings and credit associations and banks—keep apart, far reaching development processes are unlikely to result' (Seibel: 392).

The reality, of course, is not as bleak as Seibel suggests. Formal and informal financial institutions do not keep apart, nor have they in the past. In colonial periods linkages between the two have existed almost from the time that Western banks established branches in Africa and assisted the Lebanese trader and his African brokers to bring the coffee, cacao and oilpalm kernels from small family farms in the interior to warehouses on the coast. The same links were established in Asia between European banks and Chettiars and Chinese merchants and moneylenders.

There is no clear dividing line between formal and informal financial markets, the two are closely intertwined. Funds find their way back and forth between the two markets and go wherever the return is higher (Timberg & Aiyar: 279).[42] A bank loan to a big farmer may percolate to his tenants via an informal transaction, a crop loan may be spent on fertilizer after which the money moves into the informal financial system. There are almost no barriers to entry into informal financial markets: minor government officials of Tamil Nadu (Harriss, 1981: 166) and factory hands of the Kirloskar plant and sugar factories in Sangli became petty pawnbrokers when the opportunity to do so arose.

The Sangli experience demonstrates that linkages seem to develop spontaneously whenever and wherever people perceive such links as advantageous. Without such advantages, any policy of link-promotion will be an uphill struggle.

Notes

1a. ROSCA = Rotating savings and credit association, in which members' periodic contributions are pooled and given to each of them in turn.
1b. Nayar (1986: 35) has dubbed this type for India 'Lot chitty with discounts'.
2. For greater detail of the development of Maharashtra's sugar economy, the reader is referred to the many publications of Attwood and Baviskar, of which a selection is listed in the References at the end of this book.
3. Data on irrigation in Sangli are taken from Hindori and the official Socio Economic Review of the district.
4. Source: District Deputy Registrar of Co-operative Societies, Sangli.
5. The Kirloskar factory in Ramanandnagar, Tasgaon taluka, is famous for its agricultural machinery and has earned a train stop at Kirloskarwadi along the Bombay–Bangalore line.
6. Source: District Deputy Registrar of Co-operative Societies, Sangli.
7. Political interference is also reflected in the establishment of co-operative sugar factories in areas without much cane cultivation, leading to dormant factories and wastage of public money.
8. The subsidy may be greater if PACS cannot break-even financially.
9. Policy circular for UCS of the Commissioner for Co-operation and Registration of Co-operative Societies in Maharashtra, Poona 1981.
10. This was in 1984. Maximum interest rate prescribed by the RBI in 1985 was 17.5 per cent, down from 19.5 per cent in 1981.
11. Most UCS collect funds at an average rate of circa 7–8 per cent and loan them at 17–18 per cent, leaving a spread of 10 per cent. Collection agents are paid 3 per cent commission, bringing the margin to 7 per cent. The spread, however, is only 3 per cent in case the UCS has to turn to the DCCB for a loan. Such borrowing usually is only for limited amounts.
12. See map 2; Narsinhpur and Ramapur were added later.
13. This is the same practice used in commodity ROSCA reported by Bouman (1984).
14. See for example, 'The Miraculous Pawnhouse' in Newsweek 36, 1985. All major Dutch newspapers have carried pawnbroking features after a T.V. special in 1985 revealed the revival of the municipal 'Uncle Jan' in Amsterdam.
15. Pawnshops worldwide are operated both publicly and privately. The public ones are part of the formal financial sector and often run by municipal authorities. Private houses are either licensed or unlicensed. The latter are considered to operate illegally and evidently belong to the informal sector. Licensed pawnshops, which are subject to official control and interest rates, are strictly speaking not part of the informal sector. But because such control in practice exists only on paper, these shops are usually counted as part of the informal financial sector.

16. The figures of Wells and others are indicative but not comparable. Wells' figures are in percentages of total amounts borrowed, taking institutional and informal loans together. The 1986 figures are in percentages of farmers borrowing in the informal market. If the institutional loans of 1986 are included, the proportion of pawnshop loans to the total comes to 32 per cent. But then we are still comparing the ratios of loan amounts to ratios of borrowing farmers. Further, Wells reported figures for irrigated areas only, while the other figures are for rainfed and irrigated rice farming jointly. Still, the importance of pawnhouses in the informal financial market of Malaysia is undeniable.
17. The Sri Lanka study dates from 1981. At that time, a lot of brokers were of Tamil origin. Many Tamil pawnhouses have been destroyed in the riots since then. Undoubtedly, private pawnbroking in Sri Lanka has suffered a severe setback, the extent of which is yet unknown.
18. During her research, Harriss (1981: 165) made a point of calling on the local Pawnbrokers and Jewellers Association in each market town that she visited for her study.
19. The financing of moneylenders by other moneylenders, plus the fact that big lenders use brokers to bring in business (Timberg and Aiyar: 283, 285) suggest the existence of a network of informal financers not unlike the network of banks. The usual contention is that informal finance operates over only a limited geographical distance and merely shuffles around local savings. A network of informal financial intermediaries, however, greatly increases their range of action.

    The regional Chit Funds by Nayar (1973) also extend beyond the local level. 'The many chains of credit that run back from village shops to regional warehouses, to big merchant firms, to overseas capital markets' (Drake: 148) suggest a more expansive field of operation than assumed in conventional literature on informal finance.
20. One may venture several guesses to explain this apparent paradox. One is that many pawnhouses used to deal in all types of goods and gradually have become a kind of junk shop. Chinese pawnhouses in Jakarta, Indonesia, may specialize in second-hand clothes. Also, many customers of the pawnbroker belong to the poorer strata of society and borrow tiny sums of money. This typical 'poor-borrower profile' could be a second reason. Another is the moral disapproval of making money with money by charging what appears to be usurious interest rates. This is particularly the case with the unlicensed petty pawnbroker, who operates illegally on a small scale and in his own village where people know him well. Bastiaansen found that people considered his lending business highly disrespectful; sometimes even his presence at a wedding was not appreciated (Bastiaansen: 46). The licensed pawnbroker, on the other hand, is usually a respected member of the community (Bastiaansen: 33).
21. One should keep in mind that banks refinance private lenders, which increases the banks' volume of gold loans enormously.
22. RBI/NABARD circular of 1976, quoted by Bastiaansen (22).
23. Loan amounts usually equal 70 per cent of the estimated value of the pledged item; conversely, a loan of Rs 1400 implies an estimated appraisal value of Rs 2000.

24. Gramayan reports that typical bank loan applications need the support of eight or more official documents, involving ten to twelve separate visits. Processing could take between six and twelve months. In particular, farmers had to pay dearly for loans carrying a 'concessional' rate of interest. 'At every stage the farmer has to pay for fees, bribery, travel expenses, cost of refreshment for middleman and officers, as well as for his own travel and meals. He is obliged to visit banks and other offices frequently and in addition he faces loss of earnings due to absence' (Oxfam: 18). A farmer often had to leave 25 per cent of the value of his loan with the bank as margin money. These findings come from an Oxfam-commissioned survey carried out in Maharashtra and may to a large extent explain the high rates of loan default: why would a borrower repay the loan amount that he never collected?
25. The Pathans did not shun physical violence when a debt needed collection. In Sangli they built up a nasty reputation, as they did elsewhere in Southeast Asia (cf e.g. Weerasoria).
26. Still, moneylenders seemingly tend to congregate in peer groups. In Walwa subdistrict, eighteen of the nineteen registered moneylenders are Gujarati; and Khanapur subdistrict is the domain of nine reputable gold refiners in Vita town.
27. The Bombay Agricultural Debtors' Relief Act of 1939 was the first in a series of Acts meant to compel the settlement of agricultural debts. The Act was consolidated in 1947 and covered the whole State. During 1942–51, 1.5 million applications involving debts to the tune of Rs 800 million, were filed and the debts reduced to Rs 120 million. The Act also covered debts to Cooperative Societies (Seven Decades: 162). Other moneylending legislation was enacted in 1955 and subsequent years. These laws forced lenders to register with the Deputy Registrar of Co-operative Societies and to keep ledgers, notes and cashbooks in accordance with specific regulations. Premises of moneylenders became subject to inspection without prior notice. All these and other measures were ostensibly aimed at stamping out irregularities and exploitation. But the main result was inconvenience to debtors, who were suddenly deprived of a regular source of credit. After a time, the source started to flow again, often at a higher price than before, without a new and cheaper (institutional) one becoming accessible. Thus the good intentions of the Maharashtra legislators have constantly been turned towards the wrong purposes. Others have doubted even the good intentions and called the many Acts mere political gimmicks to draw votes.
28. However, later it appeared that occasionally he still transacted very large loans with a big trader or fellow-moneylender.
29. There is also a category of farmers, shopkeepers and businessmen who are not strictly wealthy, but who have sufficient means to give loans from time to time to their fellow villagers and townsmen. They mortgage loans as well as houses, accept valuables and gold as security and grant a few personal loans. Because of the extreme difficulty to get any information about their activities, they are excluded from this survey.
30. 45 loans of Rs 3000 for three months = Rs 135,000
    120 loans of Rs 1000 for six months = Rs 120,000

105 loans of Rs 1000 for one years	= Rs 105,000
30 loans of Rs 1000 for two years	= Rs 30,000
Total turnover of loans	Rs 390,000

31.
three months loans Rs 135,000/4	= Rs 33,750
six months loans Rs 120,000/2	= Rs 60,000
one year loans Rs 105,000/1	= Rs 105,000
two years loans Rs 30,000/1/2	= Rs 60,000
Average outstanding loan amount	= Rs 258,750

32. Annual inspection is carried out by the Office of the Deputy Registrar of Co-operative Societies. Charges are fixed at 1 per cent of the sum of pawn loans in the month with the highest loan sum disbursed, or Rs 500, whichever is less.
33. Although most moneylenders carry insurance, they also tend to rent a safe deposit box at the (nearest) bank.
34. Personal relationships between creditor and debtor being very important, pawnbrokers tend to offer drinks, snacks and cigarettes to their clients.
35. Even townspeople who have a stable, may keep buffalos (van den Bogaard: 46).
36. In Sangli town alone there were 158 registered milk vendors and 56 milk shops in 1985 (Ibid. 46).
37. Commercial bank loans to priority sector beneficiaries are insured under a national guarantee scheme to the extent of 75 per cent of outstanding overdues. This guarantee reinforces already existing disinterest in small-scale loans outside the purview of these priority sectors, because the insurance cover does not apply. It may also explain to some extent the lax performance of banks in collecting late loans.
38. The committee invited the Institute for Financial Management and Research, Madras, to submit a study on non-banking financial intermediaries. This study was carried out by C.P.S. Nayar, one of India's most imaginative researchers of informal finance. Nayar's extensive report of 180 pages to the Committee has been compressed into 2 pages (paragraphs 5.64 through 5.69) of the official Review Report of 370 pages and 5 pages in the Annexures to that Report. In a way, this demonstrates the official attitude to the informal market in India.
39. In a personal note to the author, Janice Jiggins remarked that recent studies from church workers in the United Kingdom, especially in areas of economic recession, indicate that informal finance markets are alive and well, there, too.
40. Bishis in India have the added advantage that they hold no interest for aspiring politicians precisely because of their small size and informal character. They are thus less tainted by favoritism and corruption.
41. Franz and Keebet von Benda-Beckmann have published extensively on the dualism between formal and informal systems in law and social security. Only part of their publications are listed in the literature references.
42. Indigenous savings and credit societies in Sierra Leone speculate in the futures market of agricultural commodities through regional traders (van Eldijk: 143–5).

References

Adams, Dale W. 1984, 'Are the arguments for cheap agricultural credit sound?' in Adams, D. W., Graham, D. H. and Von Pischke, J. D. (eds.), *Undermining rural development with cheap credit*, Westview Press.
—— 1985, 'Distinctive features of rural financial markets in Asia', *Studies in Rural Finance*, Occasional Paper no. 1203 (mimeo), Ohio State University.
Adams, D. W., Graham, D. H. and Von Pischke, J. D (eds.) 1984, *Undermining rural development with cheap credit*, Westview Press.
A.I.D. (Agency for International Development) 1985, *Small scale bank lending in developing countries*, Washington.
Annual Action Plan 1985 for Sangli District. Bank of India, Bombay.
APRACA *News Digest* vol. 9, 4. 1986.
Attwood, D. W. 1984a, 'Risk, cooperation and social mobility in Maharashtra villages', paper for the first International Conference on Maharashtra Culture and Society, University of Toronto.
—— 1984b, 'Capital and the transformation of agrarian class systems: sugar production in india', in Desai', M., Rudolph, S. H., and Rudra, A. (eds), *Agrarian Power and Agricultural Productivity in South Asia*, Oxford University Press.
Attwood, D. W. and Baviskar, B.S. (eds.), 1986, *Co-operatives and rural development*, Oxford University Press.
—— 1987, 'Why do some co-operatives work but not others? A comparative analysis of sugar co-operatives in India', *Economic and Political Weekly XXII, 26, Review of Agriculture*, A 38–55.
Bastiaansen, R. 1986, 'Pawnbroking in Sangli District', Unpubl. report (Msc. thesis), Agricultural University of Wageningen.
Baviskar, B. S. 1971, 'Co-operatives and caste in Maharashtra, a case study', in Worsley, P. M. (ed.), *Two blades of grass: rural co-operatives in agricultural modernization*, Manchester University Press.
—— 1976, *Opportunity and response, social factors in agricultural development in Maharashtra*. Institute of Development Studies Bulletin, 8, 2.
—— 1980, *The politics of development: sugar co-operatives in rural Maharashtra*, Oxford University Press, Delhi.
Benda–Beckmann, F. von, *et al.* (eds.) 1988, *Between kinship and the State: Social Security and Law in Developing countries*. Foris, Dordrecht (in print).
Benda–Beckmann, F. von 1986a, 'Islamic law and security in an Ambonese village' paper for the Symposium on Formal and Informal Social Security, Tutzing, West Germany, June.
—— 1986b, 'Social Security and legal pluralism', Paper for idem.
Benda–Beckmann, F. and K. von 1985, 'Preliminary final report on the research on law and mutual help and social security on Ambon', Hila, Ambon (unpubl.).

REFERENCES

Benda–Beckmann, K. von 1986, 'Social security and small scale enterprises in Islamic Ambon', paper for the Symposium on Formal and Informal Social Security, Tutzing, West Germany, June.

Bhaduri, A. 1982, 'The role of rural credit in agrarian reform with special reference to India', *Economic Bulletin for Asia and the Pacific*, 33, 104–11.

Bogaard, H. van den 1987, De cooperative zuivel sector in India. Unpubl. report (Msc thesis), Agricultural University of Wageningen.

Bouman, F. J. A. 'The ROSCA: financial technology of an informal savings and credit institution in developing economies', *Savings and Development*, 4, 253–76.

——— 1984, 'Informal saving and credit arrangements in developing countries: observations from Sri Lanka', in Adams, Graham and Von Pishke (eds.), *Undermining rural development with cheap credit*.

——— and Houtman 1988, 'Pawnbroking as an instrument of rural banking in the Third World', *Economic Development and Cultural Change*, October, 69–89.

Brahme, S. 1984, *Producers' Co-operatives, Experience and Lessons from India*, Institute of Social Studies, The Hague.

A Brief Note on the Working of the Sangli District Central Co-operative Bank, mimeo, 1985.

Chandavarkar, A.G. 1985a, *The informal financial sector in developing countries: analysis, evidence and policy implications*. I.M.F. (mimeo) Washington.

———1985b, 'The non-institutional financial sector in developing countries: macroeconomic implications for savings policies', *Savings and Development* IX, 2, 129–41.

Cole, D. C. and Park, Y. C. 1983, *Financial Development in Korea 1945–1978*. Canberra University Press.

Commissioner for Co-operation and Registration of Co-operative Societies in Maharashtra, 1981, *Policy Circular for Urban Credit Societies*, Poona.

Darling, M. 1947, *The Punjab Peasant in Prosperity and Debt*, 4th edn. Oxford.

Datta, B. 1982, 'Miles to Go', *Economic and Political Weekly*, 17, 34, 1361–4.

District Statistical Abstracts, Sangli.

D'Mello, L. 1980, 'Institutional aspects of lending to small farmers—the Indian case', in Howell, J. (ed.), *Borrowers and Lenders*, Overseas Development Institute, London.

Drake, P. J. 1980, *Money, finance and development*, Robertson, Oxford.

Economic and Political Weekly, 1982, Nabard's Inheritance, 17, 31, 1261.

Eldijk, A. van 1987, 'Tussen volksrecht en overheidsrecht. Over zelfregulering van marktactiviteiten in Sierra Leone', in *Recht in ontwikkeling*, Tien agrarish-rechtelijke opstellen. Kluwer, Deventer.

Esman, M. J. and Uphof N. T. 1984, *Local organizations, intermediaries in rural development*, Cornell University Press.

F. A. O. 1982, 'Chitali's Dairy, India, in *The private marketing entrepreneur and rural development*, Agricultural Services Division, bulletin no. 51, Rome.

F.A.O./O.S.U. 1986, *Bibliography on agricultural credit and rural savings*, Second series no. 1, Rome.

Franda, M. 1979, *India's rural development*. Indiana University Press.

Gamba, Ch. 1958, 'Poverty and some socio-economic aspects of hoarding, saving and borrowing in Malaya', *Malayan Economic Review*, October.

Ghate, P. B. 1986, 'Some issues for the regional study on informal credit markets', Discussion paper for the Design Workshop, Asian Development Bank, Manila, mimeo.

Gerner, H. 1985, 'Rural financial markets in Sangli District, India', Unpubl. Report, Agricultural University of Wageningen.

Government of India 1972, *Report of the Banking Commission*, Bombay.

Groot Kormelink, J. 1985, 'Agricultural credit societies in Sangli District, India', unpublished report, Agricultural University of Wageningen.

Harriss, B. 1981, *Transitional trade and rural development*, Vikas Publishing House, Delhi.

——1983, 'Money and commodities: their interaction in a rural Indian setting', in Von Pischke, J. D., Adams, D. W. and Donald, G. (eds.), *Rural financial markets in developing countries—their use and abuse*, Johns Hopkins, Baltimore.

Hill, Polly, 1982, *Dry grain farming families, Haussaland (Nigeria) and Karnataka (India) compared*, Cambridge University Press.

Hindori, M. 1986, 'Lift irrigation schemes in the Sangli District', unpublished report, Agricultural University of Wageningen.

Holst, J. U. 1985, 'The role of informal financial institutions in the mobilization of savings', in Kessler, D. and Ullmo, P. A. eds. *Savings and Development*, Economica, Paris.

Hospes, O. 1985, 'Semi-autonomus and autonomous saving and credit institutions in rural Sangli, unpublished report, Agricultural University of Wageningen.

Howell, J. (ed.) 1980, *Borrowers and Lenders*. Overseas Development Institute, London.

Khan, A. M. 1984, 'Farm credit and the Sixth Five Year Plan', *Pakistan Agriculture*, 6, 11, 36–9.

Kortenhorst, J. and K. Zevenbergen, personal letters to the author from India, February–April 1988.

Kumarasundaram, S. 1982, 'The Indian financial system. Its deficiencies and some remedies', *Economic and Political Weekly*, 17, 19, 793–8.

Kurup, T. V. Narayana 1976, 'Price of rural credit, an empirical analysis of Kerala', *Economic and Political Weekly*, 3 July, 998–1006.

Maharashtra State Co-operative Bank 1985, *Seven decades of innovative banking 1911–1981*, Bombay.

——undated, 'Progress and achievements of the Bank', mimeo, Bombay.

Maharashtra State Co-operative Land Development Bank 1985, 'Note on the working of the Bank', mimeo, Bombay.

McLeod, R. 1980, 'Dualism in financial markets', in Fox, J. J., Garnaut, R. G., McCawley, P. T. and Mackie, J. A. C. (eds.), *Indonesia: Australian perspectives*, The Australian National University.

Moll, H. A. J. (1989), *Farmers and finance, experience with institutional savings and credit in West Java*, Agricultural University of Wageningen.

Morris, F. 1985, 'India's financial system: an overview of its principal structural features', World Bank, World Bank Staff Working Paper no. 739, Washington D.C., 76 pp.

National Land Development Banks Federation 1984, *A Study on Recovery of the LDB's Dues through Credit Market Linkages (Maharashtra)*, Bombay.
———1985. *Five Year Plan 1985–1990*, Bombay.
———1985, *Co-operative Land Development Banks and Agricultural Development*, Silver Jubilee pamphlet, Bombay.
Nayar, C. P. S. 1973, *Chit Finance*, Vora & Co., Bombay.
———1982, 'Finance Corporations, an informal financial intermediary in India', *Savings and Development*, VI, 1, 5–40.
———1986, 'Can a traditional financial technology co-exist with modern financial technologies? The Indian experience', *Savings and Development*, X, 1, 31–56.
Newsweek 1988, 'Asia's gold rush', 15 August, 20–5.
Nieuwkoop, M. van 1986, 'Rural informal credit in Peninsular Malaysia', unpublished report (Msc. thesis), Agricultural University of Wageningen.
Notes and News 1985, Special column in *The Maharashtra Cooperative Quarterly*, 68, 3, 247–9.
Oxfam 1987, *A manual of credit and savings for the poor of developing countries*, Oxford.
Platteau, J. P. 1980, 'Rural credit from too far above', *Economic and Political Weekly* XV, 2, 60–2.
Platteau, J. P., Muriekan, J., Palatty, A. and Delbar, E. 1980, 'Rural credit market in a backward area. A Kerala fishing village', *Economic and Political Weekly*, Special number, October, 1765–80.
Platteau, J. P. and Abraham, A. 1984, 'Credit as an insurance mechanism in the backward rural areas of less developed countries', *Savings and Development*, VIII, 2, 115–34.
Prakash, B.A., 1984. 'Private financing firms in Kerala, A study', *Economic and Political Weekly*, XIX, 50.
Rangaswamy, A. 1974, Contractors' Opera. *Economic and Political Weekly*, 9, 5.
Reserve Bank of India 1985a, *Annual Report*, Bombay.
———1985b, *Report of the Committee to review the working of the monetary system*, Bombay.
Schlesinger, L. I. 1981, 'Agriculture and Community in Maharashtra', *Research in Economic Anthropology*, vol. 4. Jai Press, 233–74.
Seibel, H. D. 1985, 'Saving for development, a linkage model for informal and formal financial markets', *Quarterly Journal of International Agriculture*, 24, 4, 390–8.
Seibel, H. D. 1986, 'Rural Finance in Africa: The role of informal and formal financial institutions', *Development and Co-operation*, 6, 12–14.
Seibel, H. D. and Marx, M. T. 1987, *Dual Financial Markets in Africa*, Case studies of linkages between informal and formal financial institutions, Breitenbach, Saarbrücken–Fort Lauderdale.
Seven Decades of Innovative Banking 1911–1981. Maharashtra State Co-operative Bank, 1985.
Shetty, S. L. 1978, 'Performance of commercial banks since nationalisation of major banks. Promise and reality', *Economic and Political Weekly*, Special number, August, 1407–51.
Sivakumar, S. S. 1978, 'Aspects of agrarian economy in Tamil Nadu: a study of

two villages. Part III, Structure of assets and indebtedness', *Economic and Political Weekly*, 20 May, 846–51.

Socio Economic Review and District statistical abstract of Sangli District 1980-1981. Government of Maharashtra, Bombay.

Timberg, T. and Aiyar, C. V. 'Informal credit markets in India', Economic and Political Weekly, Annual Number, February, 279–302.

Vogel, R. 1984, 'Savings mobilization, the forgotten half of rural finance', in Adams, D. W. Graham, D. H. and Von Pischke, J. D. *Undermining rural development with cheap credit*, Westview Press.

Von Pischke, J. D 1981, 'The political economy of specialized farm credit institutions in low income countries, World Bank Staff Working Paper no. 446, Washington D. C.

Von Pischke, J. D. 1983a, 'Local rural financial institutions', in Von Pischke, J. D., Adams, D. W. and Donald, G. (eds.), *Rural financial markets in developing countries*. Johns Hopkins University Press, 227–31.

———1983b, 'Changing perceptions of rural financial markets', in Von Pischke *et al.* (eds.), *Rural financial markets in developing countries*, Johns Hopkins University Press, 1–13.

Wadhva, C. D. 1979, 'Rural banks for rural development: the Indian Experiment', *Contributions to Asian Studies*, XII, 179–95.

Weerasoria, W. S. 1973, *The Nattukottiai Chettiar Merchant Bankers in Ceylon*, Tisara Praasakayo Publishers, Colombo.

Wells, R. J. G. 1979, 'Sources and utilization of rural credit in Mukim Langgar', *Malayan Economic Review*, III, 2, 89–97, Malaysia.

———1980, 'The informal rural credit market in Peninsular Malaysia', unpublished report.

Appendix 1

ABBREVIATIONS AND ACRONYMS

BISHI	Marathi vernacular for local informal savings and credit club
CB	Commercial Bank
EPW	*Economic and Political Weekly*
FAO	Food and Agricultural Organisation
IRDP	Integrated Rural Development Plan
LDB	Land Development Bank
NABARD	National Bank for Agriculture and Rural Development
PACS	Primary Agricultural Credit Society
OSU	Ohio State University
RBI	Reserve Bank of India
RESCA	Regular Savings and Credit Association
ROSCA	Rotating Savings and Credit Association
RRB	Regional Rural Bank
SCB	State Co-operative Bank
TALUKA	Subdistrict
UCS	Urban Credit Society

Appendix 2

Poona Herald 3/8/80

WRITE-OFF OF LOANS

Sir,
The recent decision of Government of Maharashtra to waive the co-operative farm debts to the tune of Rs 49 crores will have serious repercussions on the attitude of villagers. It may be a vote-winning gimmick on the part of the new government, but at what cost? I think it is high time for the intelligentsia to raise their voice on a common platform rather than looking at this problem passively.

The tendency should be nipped in the bud lest it should spread to the other sectors of the economy, particularly the small-scale units, the majority of which are in doldrums. The time is not too far off when such requests may come up in respect of agricultural advances by commercial Banks. The Reserve Bank of India has, therefore, rightly put its foot down on this policy.

Such type of actions on the part of the government can even lead in loss of faith of the general public in keeping deposit in banks.

July 31   S. M. Murdeshwar

Poona Herald 4/8/80

COOP. CHIEF WARNS AGAINST WRITE-OFF-LOANS POLICY

New Delhi, Aug. 3: The National Co-operative Union of India (NCUI) President B. S. Viswanathan today strongly opposed the moves of certain State governments to write off loans extended to farmers by co-operative banks.

Referring to the proposals of Maharashtra and Tamil Nadu governments on this issue, Mr. Viswanathan said if the proposals were accepted, the farmers in other States would also start agitation demanding writing off of their short and medium-term loans.

He told newsmen here that instead of writing off loans, the State governments could subsidise the interest rates on the loans. In cases of natural calamities like drought and floods, the loan repayment of farmers could be re-scheduled or postponed, instead of writing them off, he added.

Mr. Viswanathan said if writing-off-loans was allowed now, "All election year would become writing-off-loan years".

He demanded that co-operatives should be brought under the Central or concurrent list of the Constitution to accelerate the growth and expansion of co-operative movement in the country. One reason for the uneven growth of co-operative movement in India is that now it was a State subject. The Centre was finding it difficult to get its decisions and programmes implemented at the State level.

He also wanted the Centre to bring forward a comprehensive legislation to govern the functioning of co-operative federations and multi-state co-operatives.

Mr. Vishwanathan said the Union was holding a conference of the co-operative parliamentary forum here this week to discuss various issues, including, restoration of the democratic character in the co-operative institutions. He said about 250 members of parliament belonging to various parties are expected to attend the conference.

—(UNI)

Index

Adams, Dale W., 1, 8, 10, 18, 35, 121
Agricultural credit, 7, 12
Agricultural Development Plan, 37
AID, Agency for International Development, 123
Aiyar, 8, 76–80, 87, 88, 91, 125, 127
All India Rural Credit Survey, 12
Annual Action Plan, 82, 114
Asian and Pacific Regional Agricultural Credit Association (APRACA), 9
Asian Development Bank, 9
Attwood, D. W., 24, 27, 28

Bastiaansen, R., 73, 83, 84, 89, 93, 95, 97
Baviskar, B. S., 24, 27, 28
Benda Beckmann, F. and K. von, 131 n. 41
Bhaduri, A., 119
Bishi, 2, 4, 30, 32, 41, 49, 51, 52–69
  characteristics of, 56, 58–61, 62, 65–6
  commodity bishi, 64
  comparison with UCS, 58–61, 62, 64–6, 116
  cycles of duration of, 57–60, 67
  default in, 62, 64, 68
  developments in culture of, 64, 66–9
  emergence and origin, 53, 55–7, 65, 115–16
  fines in, 57, 61–3
  grain bishi, 53
  growth and location of, 56, 59, 64–5
  interest rates in, 57, 61–7
  lending dimension of, 57–8, 60–4, 67–8
  linkages with formal finance institutions, 60, 63–4, 66, 69
  management and leadership, 59–61, 66
  membership, 57–60, 63–5, 68
  operational costs and procedures, 61–6
  purpose of loans, 57–8, 61
  raison d'être of, 56–7, 59, 115–16
  savings dimension of, 56–9, 61–4, 66–7
  types of, 52–3, 56, 63
Bogaard, H. van den, 28, 59, 105, 106, 108
Bouman, F. J. A. 8, 52, 68, 72, 75, 80, 98, 99
Brahme, S., 43

Central Bank, 5, 122; see also Reserve Bank of India
  control of, 6
Chandavarkar, A. G., 8, 99, 115, 120
Chit Fund (Chitty), 53, 55, 117
  Act, 12, 117, 120
  savings and investment and, 53
Cole, D. C., 5
Commercial Bank, 2, 7, 9, 11–18, 30, 32, 34, 45, 48
  access to, 15
  default of bank loans, 17, 114–15
  development role of, 15–17, 31, 106, 114–15
  morale of . . . staff, 17, 114–15
  nationalization of, 15
  network of, 15, 30–1
  pawnbroking by, 81–6
  viability of, 114–15, 122–4
Co-operative(s), 1, 4, 5, 7, 12, see also PACS
  . . . circular, 14
  Commissioner for co-operation, 31, 44, 61
  dairy co-operatives, 105–8, 110
  District Central Co-operative Bank, (DCCB), 30, 36–9, 83, 88
  organizational structure of, 14, 34, 39
  pawnbroking in, 82–3, 88, 89

INDEX

Co-Operative(s), (*Contd.*)
   politics and, 31, 35, 39–43, 63, 123, 128 n. 7
   Registrar of, 31, 59, 87
   ... Societies Act, 31, 44, 61
   State ... Bank, 14, 34
   sugar ... factories, see sugar
   Urban ... Bank, 30, 46, 82–3
Credit: agricultural, 6, 7,
   allocation policies, 11, 17, 37
   for dairy development, 104–12
   implications of cheap credit, 7, 16–18, 35, 43, 121
   institutional credit, see formal finance institutions
   ... and marketing, 35, 37, 42
   ... requirements, 14, 16, 18, 121
   see also Bishi; Co-operatives; Informal credit; ROSCA; subsidy; UCS

Dairy development in Sangli, 2, 4, 27–9, 103–8
   formal sector loans for, 35–6, 39, 106, 108
   informal moneylending in, 108–12
   milk production in, 104–7
Darling, M., 76
Datta, B., 17
Default(ers), 7, 14,
   of bishi loans, 62, 64, 68
   of Commercial Bank loans, 17, 114–15
   of DCCB loans, 38
   of LDB loans, 35
   of PACS loans, 40–3
   of UCS loans, 47–8, 50
   of pawnbroking loans, 99
District Central Co-operative Bank (DCCB) see Co-operative
D'Mello, L., 14, 15
Drake, P. J., 10, 71, 72, 78, 80, 121

Economic and Political Weekly (EPW), 14, 17
Eldijk, A. van, 131 n. 42
Eradication of poverty, see Poverty
Esman M. J., 46

FAO, 104, 119, 121
Finance: ... corporation, 6
   development and finance in Sangli, 113–16
   politics and finance, 15, 17, 34–43, 114
   role of finance, 1, 2
   rural finance in India, 11–19
Financial intermediaries, 1, 2
   performance of, 3
Financial markets, 2, 4
   composition of rural ... 5, 11–19
   development of in Sangli, 30–3, 124–7
   linkages in, see informal finance market
   new frontiers of technology in, 125–7
   regulation of, 17, 117–20
Financial policy in India, 11–19, 113–4
Formal financial institutions, 1, 2, 5, 11–19
   deficiencies of, 16–19, 114–7, 130 n. 24
   linkages with informal finance, see Informal finance market
   outline of ... in Sangli, 30–33
   role of ... in development of Sangli, 113–15
   see also Central Bank, Commercial Bank, Co-operatives, NABARD, RBI
Franda, M., 14, 17, 41, 48, 70, 76, 87, 89, 117

Gamba, Ch., 72, 74
Gerner, H., 57, 62
Ghate, P. B., 9, 127
Gold (and jewellery), and pawnbroking, 74–102
   as security for loans, 37, 48, 69
   hoarding of, 16, 53, 71–2
   popularity of saving in, 71–2, 93, 101
   prices of ... in India, 73, 83 (footnote)

Harriss, B., 13, 72, 76–80, 81, 87, 89, 91, 94, 127
Hill, P., 8, 18, 19, 21, 51, 77, 80, 88, 93
Holst, J., 8, 9, 125, 127
Hospes, O., 48, 49, 57, 63

Houtman, R., 72, 75, 80, 98, 99
Howell, J., 121

Informal finance market, 4–10, 12, 14, 16, 19, 32–52, 66
  advantages and disadvantages of, 8–9, 81, 116–17, 124–5
  complementarity of . . . and formal finance institutions, 2, 65, 121–5
  debate formal and . . . 5–10, 33, 117–27
  dependancy of rural households on, 2, 12, 17–19, 70, 92–3, 115–17, 121
  description of, 6
  innovations of, 125
  intermediaries of, 2–3, 6, 9, 12, 125
  linkages of . . . with formal finance institutions, 10, 69, 80, 82, 110, 112, 125–27
  networks in, 129 n. 19
  opportunity and response of, 115–17
  replacement of . . . by formal finance institutions, 5, 8, 121–5
  size of, 12
  see also bishi, milk collectors, moneylenders, pawnbrokers, UCS
Indian Banking Commission, 9
Integrated Rural Development Plan (IRDP), 15
Interest rates, 7
  of bishi, 52–69
  of the co-operative sector, 34–43
  of milk collectors, 110
  of moneylenders and pawnbrokers, 70–102
  of partnership firms, 118–19
  of UCS, 44–51
  control of by RBI, 11, 31
  perferential . . . 11, 18, 35
  subsidies and, 9, 15, 17
Irrigation, 20–4
  canal . . . in Sangli, 22–3
  lift irrigation, 23, 35–6
  loans for, 35–6, 39, 78
  from wells, 21, 23

Jaggeries, 20, 27, 28, 42
Jewellery, see gold

Karnataka State, 21
Khan, A. M., 121
Kortenhorst, J., 95
Krishna river, 21, 23
Kumarasundaram, S., **16, 17**
Kurup, T. V. N., 77, 78, 89, 91, 94

Land Development Bank (LDB), 30, 34–6
Linkages between formal informal finance, 10, 69, 80, 82, 110, 112, 125–7

Malaysian Rural Credit Survey, 74–5
McLeod, R., 5, 74, 80, 99, 126, 127
Milk collectors, 3, 58
  competition between, 109–11
  lending by, 3, 103, 108–12, 116
  purpose of loans by, 110
Moll, H., 71, 72
Monetary policy of India, 11, 115
Moneylenders, 6, 8, 14, 31, 56, 70–102; see also pawnbroking
  Acts, regulations and, 12, 77, 87–8, 118–21, 130 n. 27
  Associations of, 76, 86, 129 n. 18
  categories of, 70, 76–8, 87, 89, 109–10
  characterizations of, 14, 18–19, 33, 76, 78–80
  clients of, 70, 76–7, 87–96
  •linkages with formal finance institutions, 77, 80
  milk collectors as, 108–12, 116
  performance of, 80, 117–19
  registration of, 32, 86–7
  stereotypes of, 18–19, 33, 76, 78–80, 119–20
Morris, F., 11 12, 17 18

NABARD, National Bank for Agriculture and Rural Development, 16, 23, 34
Nayar, C. P. S., 53, 55, 72, 89, 131 n. 38
Newsweek, 73
Nieuwkoop, M. van, 72, 75, 78

Ohio State University, 8, 119, 121

Opportunity and response in dairy and sugar development, 24–9
Overdue loans, see Default
Oxfam, 80

PACS, 14, 30–1, 39–43
 bishi and, 41
 fraud and loan default in, 40–3
 growth and decline of, 14–15, 30, 42
 interest rates of... loans, 3, 9
 loan policy of, 39–40, 42
 management and leadership, 39, 41–2
 membership, 41, 43
 mobilization of savings by, 39
 official Indian policy towards, 14, 30–31, 39–41
 operational costs of, 43
 pawnbroking and, 83
 performance of, 41
 politics and, 40–3
 purpose of loans by, 39
 UCS and, 41
 viability of, 42–3
Park, Y. C., 5
Partnership (Act, Firm), 117–21
Pawnbroking, 4, 70–102; see also moneylender
 borrowing periods in, 75, 78–9, 84–102
 by formal finance institutions, 74, 81–102, 128 n. 15
 by private lenders, 70–80, 102
 clients of, 71, 76–80, 89–90, 94
 comparative advantages of, 81, 99
 economics of, 96, 97–100, 130 n. 30, 131 n. 31
 interest rates in, 74–102
 licensed pawnbroking, 87–92
 linkages between formal and informal, 72, 80, 82
 monopoly profits of, 74, 76, 79, 96–7
 nature of, 71, 89
 new entrants into, 76–7, 89, 94
 peaks in... activities, 86, 92
 popularity of in Asia, 71–80, 100–2
 preference of borrowers for private, 75, 85, 91, 95
 purpose of loans, 7–8, 86, 91–2, 95

 RBI and, 82–3
 registration of, 87–9
 repledging of pawns in, 82–3
 social acceptance of, 81, 89, 96, 129 n. 17 and n. 20
 transactions, costs of, 71, 85, 91, 99
 types of pawns, 71, 78–9, 87, 95
 unlicensed pawnbroking, 92–7
Penny economy, 8, 9, 18, 99, 122–5
Pigmy deposit scheme, 47
Platteau, J. P., 78, 79
Pledging of gold and jewellery, 37, 69, 70–101
Poverty, alleviation of, 15, 114
 eradication of, 5, 11
Prakash, B. A., 117, 118, 119
Primary Agricultural Credit Societies (PACS) see PACS
Priority sector, 11, 16, 18, 114–15

Rangaswamy, A., 27
Regional Rural Bank (RPB), 12, 15, 16, 123
Regular Savings and Credit Association (RESCA), 52–3, 55, 56
Reserve Bank of India (RBI), 2, 11–17
 attitude of... towards informal finance sector, 83, 87–8, 90, 117, 131 n. 38
 control and regulation by, 11, 31, 36–9, 42, 88, 90
 reports of, 115, 117, 120, 124
Review Committee (Report), 14, 16, 18, 41, 121, 124
Rotating Savings and Credit Associations (ROSCA), 3, 6, 8, 52, 59, 68, 117, 120, 125
 basic variations of in Sangli, 53–5
 characteristics of, 52, 53, 55, 59
 history of in Sangli, 55
Rural Banks, see Commercial Banks
Rural development in Sangli, 113–16; see also dairy, sugar
Rural finance (market), see finance

Savings: collection and mobilization of, 9, 11, 15, 16; see also bishi, PACS, UCS

INDEX

... facilities, 7
... needs, 47
Schlesinger, L. I., 21
Seibel, H. D., 127
Self help groups, 6, 9, 31, 44, 52–6, 65–6; see also bishi, ROSCA
Shetty, S. L., 16, 17, 18, 121, 123
Sivakumar, S. S., 72, 78, 79, 91
Societies, see also PACS, RESCA, ROSCA, UCS
  Co-operative ... Act., 31, 42
  indigenous savings and credit societies, 131 n. 42
  lift irrigation societies, 23
  salary earners societies, 30, 32
  village societies, 15, 31
Socio Economic Review, 21, 106
Subsidy of credit and loans, 9, 15, 17, 35, 37, 108; see also preferential interest rates
Sugar: .. belt, 23–7
  .. boom in Sangli, 2, 3, 20–9, 36, 56, 65
  .. cane cultivation, 3, 21–9, 112–13
  .. cane and dairy development, 103, 110
  effects of .. economy on demand for financial services, 28–9, 114–17
  .. factories, 20, 21, 24, 28, 35, 37, 128 n. 7
  government support for cane cultivation, 113–14
  politics and, 2, 20, 21, 128 n. 7
Syndicate Bank, 123

Timberg, T., 8, 76–80, 87, 88, 91, 125, 127

Unregulated (finance) market, 5
Unorganized (finance) market, 5
Uphof, N. T., 46
Urban Credit Societies (UCS), 4, 30–2, 41, 44–52, 56
  collection agent for, 47–8, 51, 128 n. 11
  fraud and loan default in, 47–8, 50
  growth and location, 44, 46
  interest rates of, 47–8
  loan policy of, 48, 50
  management and leadership, 46, 50–1
  membership, 44–6, 50
  mobilization of savings and capital, 45–8, 50
  official policy towards, 45, 48
  operational costs, 50
  pawnbroking and, 82–3
  performance and viability, 45–51
  pygmee deposits in, 47
  politics and, 44, 46, 49–51, 64
  raison d'être of, 44–5, 50, 115–16
  registration of, 44, 45
  relations with bishi, 51, 56, 58–61, 64, 66, 69

Vogel, R., 47
Von Pischke, J. D., 2, 7, 8, 18, 121

Wadhva, C. D., 119
Wageningen University, 3
Walwa river, 23
Wells, R. J. G., 74, 78
World Bank, 11, 12, 36

Zevenbergen, K., 95

DATE DUE			
	261-2500		Printed in USA